The Electric-Powered Bicycle Lamp 1888

by Peter W. Card BSc FIA

The John Pinkerton Memorial Publishing Fund

Following the untimely death of John Pinkerton in 2002, a proposal was made to set up a fund in his memory.

The objective of the Fund is to continue the publishing activities initiated by John Pinkerton, that is to publish historical material on the development of the bicycle of all types and related activities. This will include reprints of significant cycling journal articles, manufacturers' technical information including catalogues, parts lists, drawings and other technical information.

Published by the John Pinkerton Memorial Publishing Fund, 2006

ISBN 978-0-9552115-2-2

Printed by Quorum Print Services Ltd. Cheltenham.
December 2006

Nefa Battery Production, 1930's

CONTENTS

When the road's dark we can both despise p'liceman and lamps as well
there are bright lights in the dazzling eyes of beautiful Daisy Bell.

From the song 'Daisy Bell' by Harry Dacre, composed in 1892 and sung by Katie Lawrence.

Publications By The John Pinkerton Memorial Printing Fund

Lightweight Cycle Catalogue Volume 1

An Encyclopaedia of Cycle Manufacturers - compiled by Ray Miller

Frederick H Pratt and Sons -Complete Cycle Engineers - Alvin J E Smith

The Electric-Powered Bicycle Lamp 1888-1948 - Peter W. Card

All publications are available through the Veteran-Cycle Club sale officer.

FOREWORD

Foreword by Derek Roberts.

November 2006

There are few left who remember cycling in the twenties and thirties of the last century, but they probably remember without affection the lamps available for night riding. These ranged from the few remaining candle lamps and cheap bobby-dodgers to the expensive lamps such as the outsized and heavy acetylene gas-powered 'Calcia King'. Most of the oil lamps used paraffin, sometimes blended with colza, but a minority used the dearer special lamp-oil which produced less soot. Unfortunately, even the best oil lamps were not all that effective for illuminating the road ahead, and the small tin-plate bobby-dodgers were almost useless. Although electric lighting had been in existence for many years, there were no cheap, reliable electric lamps available until the thirties. Dynamo lighting had been available earlier, but was probably too expensive for many cyclists, who were largely an impecunious lot. Nevertheless, some of today's survivors will talk longingly of the light provided by the big acetylene lamps, and will assert emphatically that the light they gave was much superior to all other forms of illumination.

The history of these early methods of lighting has already been described by Peter W. Card in his numerous writings [Joseph Lucas 'The First King of the Road', Early Vehicle Lighting (Shire) and others] and he has now performed a similar service for electric bicycle lamps. In this, his latest book, The Electric-Powered Bicycle Lamp 1888-1948, he starts with the discovery of electricity by Humphrey Davy in 1809 and continues with an outline of the work of other scientists, whose work on electric power for domestic lighting preceded Joseph Lucas's production of the first battery-powered bicycle lamp in 1888.

With hindsight we can see that it would only be a matter of time before non-rechargeable batteries would appear, and very early examples of dynamo lighting would be introduced before the Great War. Strange as it may seem to us today, many riders were reluctant to abandon their 'cheap and smelly' oil lamps, but possibly cheapness triumphed over efficiency. A real breakthrough was made when Miller & Co produced a cheap and efficient dynamo lamp. Joseph Lucas Ltd were slow to reintroduce a battery-powered lamp, but of course it was only a matter of time before they came to consider that their range of oil lamps was inadequate for the needs of cyclists, even though many of them were loath to scrap their existing lamps, which were adequate for complying with the law and were cheap to run. The advocates of acetylene lamps continued to claim that the large gas-powered lamps far surpassed the newfangled battery-powered lamps, most of which were of little use as real illuminators.

With an easy-to-read text and reproductions of some significant catalogues and brochures, Peter elegantly steers the reader through the history of the medium, culminating in 1948. By then, there were only two British firms of stature selling dynamo and battery lamps. Many old clubmen still remained faithful to oil and acetylene gas. Some continental manufacturers produced oil lamps to cater for this vanishing market, but in Great Britain electricity reigned supreme.

Peter's catalogue collection was used as a basis for two highly successful publications issued by John Pinkerton in the 1980s, and it is fitting that this wholly new book is published in memory of John's name. I feel certain that collectors of the medium and researchers will find this important new work enjoyable and informative.

Acknowledgements:

The author offers his grateful thanks to Derek Roberts for his foreword; also particular thanks go to Ray Miller for his sterling efforts in helping me with much scanning and his stamina in coping with my passion, Ian Smith for proofing the text and correcting my split infinitives and Brian Hayward for setting the book so expertly. Thanks are also offered to those friends that willingly allowed me access to their collections which include: David N. Card, Christopher Brooks, Dorothy Pinkerton, Robert Cordon Champ and Andrew Heaps. I am indebted to the members of the J.P.M.P.F. committee for their advice and goodwill.

Fig 1. An H. Miller & Co advertisment for (1936) showing a feature that was common in the 1930's, a dual bulb system where the lower amperage bulb could be swithed on when transversing illuminated streets.

Introduction.

This volume is offered not as a definitive work covering the development and introduction of electric-powered bicycle lighting but rather as an introduction to the enthusiast, majoring on some of the manufacturing highlights of the medium between 1888 and 1948. The promotion of electric lighting for bicycles was significantly enhanced by insistence on the part of the manufacturers that it was safe, clean, reliable and cheap to use. That it was clean is undoubted; for the most part it was safe; however, particularly in the early days, it was neither cheap nor reliable. The Achilles' heel of the lamp before about 1910 was the bulb: the bulbs were simply not able to cope with the rigours of poor road surfaces, changes in temperature and careless handling, and replacement bulbs were expensive to buy.

The copywriters often repeated the notion that electricity 'costs you nothing', but they habitually failed to remind the buying public that the lamps themselves were an expensive commodity to purchase in the first place, that battery lamps needed a new battery occasionally and dynamo lamps needed oiling. In common with the resistance of today's multi-national oil companies to investigate the development of a non-oil-based-powered motor-car engine because of the perceived adverse effect on their profits, lamp manufacturers at the beginning of the 20th Century were certainly aware that they made much more money through sales of carbide of calcium and burning oil to the purchaser of their lamps than they would ever make out of the sale of the lamps themselves.

The apparent reluctance on the part of bicycle lamp manufacturers to invest in research and development of the

Fig 2. Melitilite advertisment for 1908

electric bicycle lamp to a point where few varieties of these lamps were offered until the mid-1920s is not easily explained. Those lamps that were offered before the Great War were technically interesting but they did not sell in high numbers.

One can tell that most manufactures lived a hand-to-mouth existence simply by looking at their petit advertisements in the cycling press and the little interest they showed in marketing their product at national cycle shows. Undoubtedly the uninformed cycling public must have also played their negative part, for they grew up with and understood oil-powered lamps. They may sometimes have been messy and difficult to keep alight in windy conditions, but people were comfortable with what they knew. The enthusiastic clubman delighted in his acetylene gas lamp, and did not seem to mind the irritating need to clean the lamp after every evening ride, for this was all part of the lore of cycling, and this sort of requirement was probably also believed to be character-building. The namby-pamby new-fangled electric lighting systems were not for them.

Notwithstanding high inflation and unemployment after the Great War, there was, come the early 1920s, a fresh new world: electric lighting in home and workshop was becoming the norm, the middle classes were more mobile and an exciting world of mysteries lay ahead, including the electric-powered bicycle lamp.

Early Developments.

Nowadays electric-powered bicycle lighting is somewhat taken for granted. We accept that electric-powered front and rear lamps are inexpensively available in battery and dynamo mediums, with durable plastic-bodied lamps easily obtainable from the high street or on the internet. We also have faith that the lamp will work when switched on, will give service for a good many years, and that the light emitted will illuminate the road ahead, ensuring clear visibility in all weathers and conditions.

It was not always so. The principles of electricity had been part of the science landscape since 1809, when Humphrey Davy, an English chemist, invented the first electric lamp by connecting two wires to a battery and attaching a charcoal strip between the other two ends of the wires, the glow of the charged carbon creating the first arc lamp.

A giant leap forward in the use of electricity for lighting purposes occurred in February 1879, when Joseph Swan demonstrated the first illuminated light bulb to the Newcastle Chemical Society, to the astonished gasps of the assembled worthies. In October of the same year, the reinvention of the incandescent light bulb by Thomas Edison in America, and his subsequent installation of the first commercial lighting system on the streets of lower Manhattan in 1882, proved the efficiency and, importantly, the cleanliness of electric lighting to the masses.

Battery Power.

The alkaline dry-cell battery is commonplace today, but in the late Victorian period the batteries were wet-cell. These batteries were actually a combination of cells, creating a device that could store and produce electricity from a chemical reaction. To do so, a battery must consist of two or more cells: that of a negative electrode and a matching positive united together with a liquid electrolyte to instigate a chemical reaction between the cells. From 1866, several variations of battery types were in regular use with telegraph equipment, and the

Fig 3 An example of Lucas's first battery powered lamp, in remarkably good condition and still retaining its original 'Bunsen' battery, the on/off switch can be seen on the side and the hinged lid for accessing the battery terminals for recharging.

most common of these were created from cell plates of zinc and manganese immersed in an ammonium chloride solution. Later, the most familiar of the two-fluid cells were the 'Daniell' type with copper plates and saturated copper sulphate. Also in use were the 'Grove' type with its platinum plates and nitric acid and the 'Bunsen' type with a negative carbon plate and a positive amalgamated zinc plate with diluted sulphuric acid as an excitant. This latter type was the most successful for small applications, since its electric current, once started, would be almost constant for about four hours.

The First Battery-Powered Bicycle Lamp.

Joseph Lucas and Sons consisted for the most part of a father and son duo, Harry the son, who looked after the day-to-day running of the Little King Street factory in Birmingham, and the Father, Joseph, who acted as a journeyman (representative) for the fledgling company. Always on the lookout for new and exciting ideas, a Birmingham jeweller named Vaughton supplied a design for an electric lamp which became the first rechargeable Bunsen battery electric bicycle lamp. (Fig 3). A patent abridgement was accepted on 27th October 1887, Numbered 14,622 with a complete specification several months later. It is likely that Vaughton was a member of the same cycling club as Harry, or at least known in Harry's bicycle club circles, because in the advertising for the lamp it is related how the inventor had 'used his prototype for more than a year under the watchful eye of the manufacturer'. The 'New Patent Electric' model first appeared in Lucas's February 1888 catalogue priced at an astonishing 55/- shillings, representing about £180 today. The high cost was attributed to the cost of the bulb and the fact that an agreed royalty payment of 6/- shillings was paid to Vaughton for every lamp sold. The lamp was only listed for about a year, and one suspects that few were sold, albeit at least two have survived. As can be seen from the illustration, the lamp was large and sturdily made, the sprung rear bracket being a copy of that used on conventional Lucas

Safety Lamps of the period. The quality tin casing sported a hinged lid for access to the 'Accumulator' charging terminals, and also to a spun brass bulb-holder with a threaded-fitting to allow for bulb replacement. Harry Lucas kept a book where he painstakingly listed items that the company wished to sell, costing out the individual component parts and labour involved. In Harry's own hand and dated 31st October 1887 it carefully lists: 'Vulcanite Cell' 1/6, 'Lead (in fact carbon) Plates' 1/- , 'Cone Front' 2/-, etcetera.

Weighing in at 4½lbs, the 'New Patent Electric Lamp' continued to be offered throughout 1888. A further Lucas catalogue was issued, I believe, just after the Stanley Show in November 1888, in anticipation of the 1889 season. On page three of this catalogue it states 'Our Electric Lamp (is) Improved and Reduced in Price' and claims that 'in its improved form the lamp is admirable for use by those who have access to convenient means of charging'. In this statement probably lies the reason for the limited success of the lamp. The lamp battery had to be charged. A source of charging was originally suggested in the February 1888 catalogue as a dynamo, but the manner of dynamo installation for the purpose is not discussed. In the November 1888 catalogue Harry states (Harry certainly wrote the catalogue text) that 'the four lamps on display at the Stanley Show were charged from a four-cell Bunsen Battery' but again, does not tell the reader how the Bunsen Battery was charged! All in all, this was a good idea presented at the wrong time and, interestingly, it would be a full 40 years before Lucas would offer an electric-powered bicycle lamp again .

Fig 4. A German manufactured Eagle Head novelty head lamp, display mounted.

Fig 5. A Miller "Bilite" battery lamp and its delivery box

Further Developments.

The Arabian Oil Co of London, introduced their 'Arabian Electric Cycle Lamp' at the end of 1893. (Fig 6. & Page 23) The manager of the firm at this time was Fred Cassell, who later went on to establish his own cycling accessories company, managed the Deptford factory and was the driving force behind the invention. The lamp was certainly marketed until 1897, and its success, although one suspects that this was somewhat limited, was perhaps partly due to its simple construction. It consisted of a square tin-plated box seated in a metal frame with a screw-down lid.

Fig 6. Advertisment for the Arabian Electric Cycle Lamp, 1896

Attached to the lid were the alternate cell plates, in this case carbon and zinc, either of which was 'easily replaceable'. The chemical reaction medium was what was referred to by the company as 'Arabian Electric Fluid', which in reality would simply have been a chloride solution, and in the 'Directions' for the 1896 model, it was suggested that a third of a teaspoon of sulphuric acid should be added to each cell, 'to brighten the light'.

Priced at 18/6 in 1896, (£45 in 2006), a fully-charged battery would last 'between 7 and 4 hours' which, given the usual propensity for Victorians to exaggerate, probably meant that a four-volt illumination lasted for a maximum of 4 hours. As well as the battery box with two collars, which allowed it to be fitted to the cross-bar of a bicycle, there was also the front lamp housing. This consisted of a backing plate with a simple switch fitted to the rear, a bulb, described by the company as an 'incandescent globe', and a front Plano lens, which possessed a bayonet-ring to hold the whole unit together. Spare bulbs were priced at 2/- (£5 in 2006).

The Marketplace.

Let us dwell for a moment and ask the question, to whom did these expensive lamps appeal? One can only assume that, within the highly-charged enthusiasm for cycling during the last lustrum of the nineteenth century, there were many cyclists who could afford such lamps. Those that could would have been what we would term today professional men, such as doctors, solicitors, accountants and the gentry - the very people who would soon abandon cycling and adopt the use of the motor bicycle and motor car, simply because they could afford to be amused by such things. I have no doubt that like candle lighting; by the cleanliness of the electric medium must have appealed to lady cyclists. They would have been attracted by the claims as to its simplicity, ease of use and efficiency, particularly in connection with their perambulations in large areas of recreation such as Hyde Park. I suspect however, that once the need for lamp servicing and the dexterity required in dealing with the technicalities of recharging had been addressed, the lamp would be quietly cast aside and the candle lamp, simple and clean, would replace the electric lamp as the favoured means of illumination.

Dynamo Power.

Dynamo-powered electric lighting was successfully introduced in about 1895 and descriptive names such as 'Dynolite' and 'Voltalite' appeared before 1900. The latter, in various versions, continued in production until the 1920s. However, according to a drawing once on display in the London Science Museum, a patent was applied for by Richard Weber of Leipzig for a dynamo to fit on the front fork of a high bicycle (Ordinary) in 1883. Ernest Tilmann and Charles Lexow entered the scene early and patented their dynamo-powered lamp, the 'Magneto-Electric Machine' in New York on 17th July 1894. The dynamo was attached to the rear of the saddle support and lowered onto the top of the tyre for its motive power, exactly where a mudguard should be, with an electric cable running along the top tube to the headlamp at the front. It is not known if it ever went into production. On the other hand, 'The Farnham' electric bicycle lamp 'that burns when the wheel is in motion'

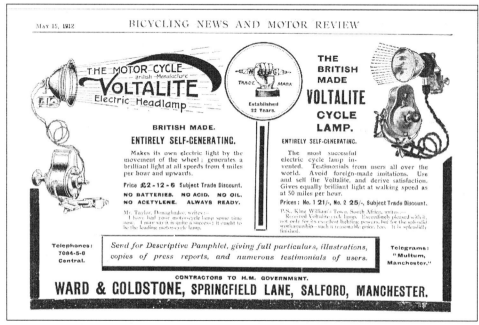

Fig 7. An advertisement for the Voltalite dynamo-powered lamp.1912.

was manufactured by Farnham Electric Co of Chicago, but in this design the dynamo was driven by the top of the front wheel with its securing bracket attached to the headstock, just above the forks. Advertised in 1898, it seems not to have survived close scrutiny and sales were minimal.

A variant on the aforementioned single headlight and separately mounted dynamo was the 'self-contained lamp' where the bulb-holder and dynamo were incorporated as one unit. The 'Pilot' was manufactured by Dowd Electric Co of Boston (USA) from 1898. Fitted to the front forks, the front wheel turned the dynamo: 'Every turn of the wheel produces light......when the machine is moving even at the

slowest rate a bright steady light is produced'. Priced at $5, several examples exist today, so one must assume that the design was reasonable successful.

The European Scene.

From a clubman's point of view, the increasing popularity of the newly-invented acetylene gas-powered lamps eclipsed the development of the electric lamp, not least because of the expense of manufacturing the delicate electric bulb and its need for regular replacement. However, once the professional riders realised that claims by the acetylene gas lamp makers regarding power and reliability were exaggerated, a gradual increase in the demand for electric lighting continued, particularly for cheaper, foreign-manufactured goods. In Europe, various companies were starting to offer battery-powered lighting. Hartzendorff & Lehmann of Berlin offered 'the newest thing', the 'Miniture Komet', as a prize item among others advertised under the

Fig 8. An example of White's petit battery lamp. A second pattern, circa 1898 version, while the original battery has long since disappeared, it retains its original delicate bulb.

Further Developments.

heading, 'Electrische Fahrrad-Lanterne'. This was a candle lamp housing with a battery fitted in a leather bag attached to the handlebars. Benjamin. B. Hoffman patented a similar devise in America on 12th May 1896, but while the detail of manufacture and its component parts were laid out in great detail, no mention is made of how to keep the battery charged.

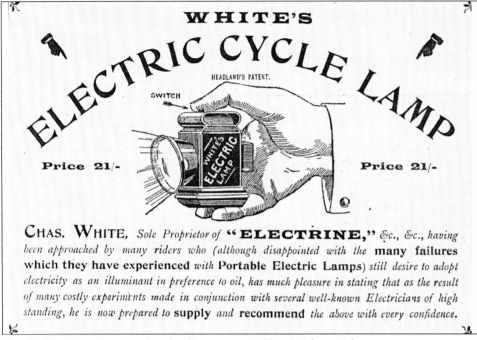

Fig 9. An advertisement for the first pattern White's electric lamp.

Charles White, the English manufacturer of 'Electrine' lamp burning oil, borrowed Headland's patent for the self-contained design of his electric lamp in 1897 (Fig 8.)and, as in the case of the Vaughton-Lucas agreement in 1888, managed to convince Henry Salsbury of the Salsbury Lamp Works, Long Acre, London, to manufacture it. Priced at 25 shillings (£54.50 today), several petit styles were manufactured, allowing only a small battery to be incorporated. This probably meant that the lighting duration was short, although it was claimed that the lamp's battery gave 'an adequate light for 10 hours'. This electric lamp's major selling point was that it was clean and small and 'in case of breakage to glow lamps (sic), duplicates can be obtained in small protecting cases to be carried in pocket or wallet'. The lamp was described as being 7 candle-power, a very doubtful claim in the view of the

writer. However, one can only speculate that it was reasonably successful because several examples are known to have survived.

The Ever Ready Company was a pioneer in the field, originally manufacturing lamps under the auspices of the Manhattan Electrical Supply Company of New York. The London company was formed in 1901 as the American Electric Novelty & Manufacturing Company. The firm, based in Charing Cross Road, imported American-manufactured goods that required cylindrical dry cell batteries also named 'Ever Ready' and similar to the modern A1 Alkaline batteries. Beginning as an independent branch of the New York based company, the UK company was re-floated as the British Ever Ready Electrical Company in 1906 and, in 1913, as The British Ever Ready Company Limited. With its trade for small, simply-made battery bicycle lamps increasing, and the exporting of lamps to Holland and Germany, from 1908 cycle accessory companies like Mead and Halford sold Ever-Ready lamps in four models. Basically they all used the same body, but with either a single convex Plano (Bulls-eye) lens for the then expensive sum of 10 shillings (a spare bulb costing 2/6), or with a plain lens at a little less. Among the most unusual and highly collectable 'Ever Ready' lamps today are those manufactured from wood. (Fig 10 & 11) With their finely cut tenon joints, chamfered corners and nickel-plated metal parts, these lamps would not earn a prize for

Fig 10. An Ever Ready wooden box battery bicycle lamp, designed to hold a large 'door bell' battery.

their aesthetics, but for simplicity, ease of use and reliability, they won much support from cyclists, and were stocked at prestigious London stores like Selfridges and Harrods. With the introduction of broadcasting and the need for High Tension Radio Batteries from 1922, the survival of the company was assured, and it continues to manufacture lamps to this day. (Fig 12.) Interestingly, inspection of the images in Ever Ready brochures for the 1930s demonstrates that the bodies of some of the lamps were certainly of Lucas manufacture, but with minor differences.

ELECTRIC CYCLE LAMP.

No. 2036.

To light the lamp, screw down the switch by turning in a clock-wise direction.

To take out a used battery, spring the handle off and remove the top cap. Ever Ready refill No. 800 should be used with this lamp, and should be inserted with the centre brass strip in contact with the bulb.

To fit a new bulb, remove front rim and reflector, then unscrew locking ring behind the bulb. The bulb and locking ring can then be easily removed together. A new bulb, Ever Ready No. 1428, should first be screwed into the locking ring and then both inserted into the lamp together.

Focussing should be effected by screwing the bulb in or out, test with the reflector in position to obtain the desired effect. When the correct position is obtained, the locking ring should be tightened.

When the lamp is in use on a cycle, the handle should be clipped behind the bracket spring to avoid rattle.

THE EVER READY CO. (Great Britain), LTD.

Fig 12. Ever Ready 1936 Advertisment

Fig 11. An Ever Ready wooden box battery bicycle lamp, designed to hold a double cell battery.

The American Scene.

In pre-1900 America, the Electric Portable Lamp Co produced what was probably the most peculiar electric lamp design in 1896. 'The Chloride' was a circular hard rubber container with a non-spillable screw lid, in which were placed zinc plates and water-diluted 'electric salt'. (Fig 13.) Discussed and illustrated in the May 1896 issue of Scientific American, the article explains that, 'at the end of the ride the contents are poured out of the lamp, and the zinc elements are carefully washed in water'. One can only assume that, with such a cumbersome method of use, the inventor had well and truly lost his way while designing this lamp!

Fig. 1.—NEW ELECTRIC BICYCLE LAMP.

Fig 13. An advertisement for the Chloride wet-battery powered electric lamp.

ECLIPSE ELECTRIC LAMP.

This electric lamp is without doubt the best of its class that has yet been brought out for the use of cyclists. It will throw a brilliant light 75 to 100 feet ahead of the wheel, and the light cannot possibly be jarred or put out unless turned out by the rider. It is as readily attached to a wheel as an oil lamp and no more difficult to operate. The current is provided by a material put up in small cans to fit into a tool bag, and is in the shape of dry powder and known as Eclipse electric sand. It costs no more than oil.

Fig 14. An advertisment for the Eclipse Electric Lamp

Wet-cell batteries were still included in designs at this time, and advertising for the Featherstone & Co 'Electra', priced at $3.50, advised that, 'the battery can be charged from an incandescent lamp or primary batteries'. A description of the lamp in The Cycle Age and Trade Review stated that, 'the battery is contained in a sheet metal case and is of the wet variety, the solution used being one part sulphuric acid to five parts water.....can be secured at any drug store'. The United States Battery Company produced their simple cylinder style U.S.B. lamp (fig 16.)which was re-chargeable 'from any direct current by using our simple re-charging device' and imaging, in the advertisement, two wires loosely pushed into a domestic bulb socket! Also in America, two box-style lamps, the 1896 Eclipse from Buffalo New York (Fig 14.)and the 1898 Acme Electric Lamp Company 'Acme' lamp, (Fig 17.) were well advertised but, like British manufacturers, the companies did not deem it necessary to linger on the method of re-charging the batteries in their advertising brochures. On the other hand, The Portable Electric Light Co of Chicago produced their 'La Marquette' in 1898 comprising an adjustable headlight and a top tube mounted cylinder to carry 'enough power to generate a light for about twenty hours' and used the new dry-cell batteries.

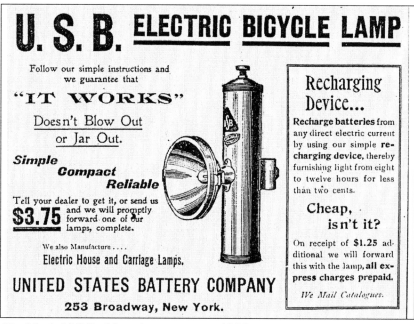

Fig 16. A, U.S.B. Advertisement circa 1900

Fig 15. A circa 1900 dynamo or battery powered lamp of unknown manufacture.

Fig17. An advertisement for the 1898 Acme battery electric lamp.

Further Variations.

A Manchester company, Ward and Gladstone, marketed their reliable Voltalite dynamo set from 1905, (Fig 7 & 18) although a less efficient model had been available from 1897. A detailed review of their 'magneto-electric' lamp appeared in Cycling magazine in 1914, claiming that the Prince of Wales possessed an example, although the article does not state if he had used it. Given the esteem generated by the royal family before the Great War and the number of times one manufacturer or another advertised that a member of the Royal Family owned an example of their manufacture, the author muses that the Royal Bicycle Shed must have been awash with lamps. The lamp was in production until 1922 although, given post-Great War inflation, it had by then doubled in price.

In 1906, GEC were the first to patent a method of making tungsten filaments for use in incandescent light bulbs; although the new filaments were costly, they effectively more than double the life of the bulb, so reducing the overall cost. By 1910, William Coolidge invented an improved method of making tungsten filaments, which subsequently proved very successful and more economical, and this is still the formula used in most domestic bulbs today. Up to this time, various methods were used to secure the bulb into the lamp. Lucas used the threaded shaft principle in 1888, but the Americans tended to favour bayonet fitting and the Europeans flange collar and, in some instances, simply push-fit.

In 1922, the Montil Manufacturing Company of Birmingham, who had previously manufactured acetylene gas lamps, believing that electric lighting was the medium of the future, produced their Bulli-Montil electric dynamo set, comprised of a headlight, wheel-rim-driven direct-current-dynamo and rear lamp. The product is interesting because it was one of the first electric lighting sets to be offered with a rear lamp, although it was not obligatory at this time to show a red lamp on the rear of a bicycle in every county of England and Wales. Another feature designed to promote sales of the combined unit was the use of the slick phrase, 'Lighting that costs you Nothing' in the company's advertising. However, they failed to remind interested parties that the lamp set required an original outlay of 35/- (£52 today) as well as the fact that, although bulbs were now more reliable and not so expensive as the pre-war types, they nevertheless needed to be replaced from time to time. The copywriters also quietly omitted to mention that the dynamo would need maintenance at some stage. The lamps were probably manufactured in Germany and assembled in England.

Fig 18. A circa 1910 Voltalite self-contained dynamo lamp, as fitted to the front forks of a 1905 Centaur bicycle. Note that the lamp is in the 'on' position with its rubber driving wheel touching the inside of the wheel rim.

Fig 19. French manufactured, nickel-plating on tin battery lamp from the mid 1920s.

Enter H. Miller & Co.

H. Miller & Co had been manufacturing bicycle lamps in Birmingham since the mid-1880s. Their post-Great War advertising seems to have been an attempt to re-write history, for they state in their 1923 brochure that they had 'started making lamps 50 years ago', which would mean that their first bicycle lamps appeared in 1873! Although this is patently untrue, the company are nevertheless to be congratulated on their industry. Miller's first foray into the manufacture of dynamo electric cycle lighting was towards the end of 1922. In their 1923 catalogue, they announced 'in no previous Autumn in history has the need for real efficiency and reliability in Cycle Lamps been greater having never had to contend with an increasing number of motor vehicles and the need for more efficient lamps'.

Fig 20. H Miller and Co circa 1936

where in the 1923 brochure is there any mention of the need for spare bulbs or maintenance, the text merely stating that the components were ready for fitting to the bicycle, 'an operation which can be performed in a very few minutes'. Clearly, Miller's were unsure of the public's acceptance of the new medium for, listed in the same catalogue, were seven acetylene gas lamps, seven oil lamps and two candle lamps.

Bicycle electric lighting was slow to become popular, but on 1st May 1929 H. Miller & Co announced a new style dynamo, this time at 21/- (£34 in 2006) for the three-component set. At the same time, two battery lamps were added to the range, consisting of a lamp using a similar body-style to the dynamo set, with bulb and two-cell dry battery, priced at 7/-, and a cheaper version priced at 3/3. Bulbs were quoted as 4-volt 3-amp priced at 10 (old) pence, and batteries were priced at 1/-, with no mention being made of the number a lighting hours that the battery would provide. Significantly, seven acetylene gas lamps were still being offered, but only five oil-powered units. By 1934, ten years after the introduction of the medium by H. Miller & Co, it was a different story. Three new designs of dynamo were offered, eleven battery sets listed at various prices between 11/6 and 2/-. To keep the more traditional riders happy, four acetylene gas lamps and five oil lamps were squeezed into the rear pages.

A headlamp, rear lamp and tyre-edge-driven dynamo sold at a cost of £1-5-0d (£39 in 2006) but, as with the products of other manufacturers, customers did not have to purchase the rear lamp if it was considered to be un-necessary. In Great Britain in 1923, Section 85 of the 1888 Local Government Act was still in force, and there was no legal requirement to carry a rear lamp. It is also noticeable that no-

Powell & Hanmer Developments.

Powell and Hanmer started manufacturing bicycle oil lamps in 1884 and, after a modest start, converted to a public company in 1902. The business partnership of Francis Powell and Frank Hanmer was successful, but there was a parting of the ways in 1910, with Hanmer purchasing Powell's shares and the latter moving on to other business interests. The interesting reason for the break-up of the partnership was that Francis Powell wanted to spend money on the development of electric lighting, while Frank Hanmer considered that it had only limited use. At this time, too, business was not at all buoyant, and P&H were only just scraping by. However, with looming war clouds, the fortunes of P&H were about to take a turn for the better.

In 1914, the War Department was not convinced that the new-fangled electric motor lighting, which was currently being espoused by a number of Birmingham lamp manufacturers, would survive the conflicts on the French terrain, and turned to P&H for acetylene and oil lamps for most of its war-time transportation lighting needs. This gave a significant financial boost to the ailing company and helped prepare them for the future. By the 1920s, high inflation and difficult trading meant that the development of electricity for motor vehicles was slow but, by 1922, a full selection of electric-powered motor lighting was cautiously placed on the market. However, another five years were to pass before P&H introduced their battery-dynamo combination 'Cycle Lighting Set'. The 1927 product was an inspired design, possibly unique to the marketplace at the time, and was comprised of a front-fork-mounted dynamo, rear lamp and a combi-

Fig 21. Powell and Hanmer front cover of their 1934 catalogue. Interestingly the only clue to the Lucas ownership of the company is the list of Lucas Ltd national depots

nation headlamp and spare battery-holder. Fitted with a change-over switch on the rear of the lamp, provision was made inside the lamp for carrying a 3.5 volt battery; 'by means of the switch the rider can change over from dynamo to the battery when the machine is stationary in traffic hold-ups'. Priced at £1/1/0d, (£34 in 2006) the set was somewhat expensive. Two battery-powered headlamps were also offered at this time, priced at a more realistic 3/6 and 5/-, (£5.50 & £8 in 2006).

In 1929, the Lucas Company purchased Powell and Hanmer for £250,000 (£8.3m in 2006), although some contemporary observers described the move as a 'Take-Over'. At first, there was little in the way of evidence for the public at large to connect the two companies; indeed, by 1930, cyclists believed that P&H had independently taken a retrograde step in electric lamp design. Side glasses on oil-powered bicycle lamps during Victorian times were necessary to identify your position to side-approaching traffic on very dark roads and, although not strictly necessary, even after the Great War the public still expected them to be fitted to acetylene and oil lamps. With the introduction of bicycle electric lighting in the 1920s, manufacturers tended to ignore this feature in their designs, together with the cumbersome rear-mounted parallel sprung brackets. However, in 1930, P&H decided to introduce not only side jewels but also sprung brackets into the design of two of their electric lamps and continued to offer models of this type well into the mid-thirties.

The Return of Joseph Lucas Ltd.

Like their competitors, the Lucas company was slow to design and market electric lighting for bicycles. In October 1927, the company offered its first dynamo-powered set, comprised of a tyre-driven generator that could be fitted to the front or rear forks, a focusable front lamp and a rear lamp. The following year saw the introduction of a battery lamp using the 'Lucas Dry Battery' and a 2.5 volt bulb. By November 1934, the Lucas Company was offering three different dynamo sets and nine battery headlamps. Amongst the latter were two new units that would set the standard of quality and clever design for many years to come. The first of these was the Lucas Twin Lamp Set. (See Fig 25) For the first time a double headlamp was designed for bicycle use, 'hitherto only motorists have enjoyed – twin headlamps with a dim and switch arrangement'. Unusually, but quite correctly, the pair of lamps possessed a reeded glass on the near-side and frosted glass on the off-side. To power the lamps, a very large battery box was mounted on the steering tube nearest the rider. Although not mentioned, the writer suspects that the unit was intended for use on tandems and tandem tricycles. Priced at 15/- for the ebony and chromium-plated version, an ivory-white-finished set was available at 16/6, (£31 in 2006). The other lamp introduced in 1934, which finds much favour today, is the imposing model 301. Spherical in shape and chromium-plated, the 5.5 volt bulb could be dimmed 'when driving in towns or well-lighted roads'. Both lamps continued to be manufactured until 1939.

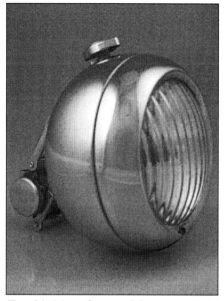

Fig 22 Manufactured between 1934 and 1937, a No 301 De-Luxe lamp using a twin-cell batteries and a dimming switch on the top for town use.

After World War Two, electric lamps were almost the only lighting medium available to cyclists. Inexpensive and lightweight oil and candle lamps continued to be manufactured on the continent throughout the 1950s, but the variety and number of models offered diminished to a point where electric bicycle lighting eventually reigned supreme. Immediate post-war models reflected those designs that were offered before 1940 albeit the grade of pressed sheet-tin was rather thinner, but the bonus at this time was the benefit of chemical research during the war years, the result being more robust filament bulbs and batteries. With the need to get out and about after the restrictions of war, there was renewed interest in cycling, particularly by family groups. One supposes that there was good business to be had by selling accessories in bicycle shops, particularly when cyclists discovered that, having put their battery lamps away on the outbreak of war, the dry-cell batteries had often oozed their contents and the resultant acid had eaten the body of the lamps, making a very nasty mess. With the advent of lighter and more efficient batteries, halogen bulbs and plastic bodies, it is rather difficult today to understand why electric lighting took so long to get established. Indeed, the development of cycling as a sport has continued to inspire modern lightweight lighting systems for 24-hour racing. Costing £650 in 2006, a 'Lupine Edison 10' using Lithium battery technology can supply a continuous 40-watt beam for nine hours; what would Harry Lucas have thought?

An electric lamp.

(54) On the stand occupied by Herbert Smith & Co. Mr. E. A. McLachlan displayed an interesting novelty in the form of a storage battery electric lamp to burn for the space of four hours without re-charging. The weight of the lamp has been considerably reduced since last shown, and now only scales 2¾ lbs. In appearance it is quite like the ordinary lamp, as our readers will see from the accompanying sketch. The price of it is 25s.

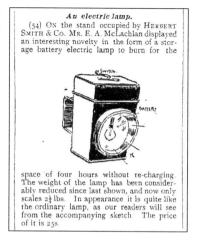

Fig 23. An editorial comment for a Herbert Smith & Co battery lamp at the Stanley Show in 1892 Lucas Twin Lamp Set, 1934.

Fig 24 .Lucas Twin Lamp Set, 1934.

It pleases me to know that collectors continue to take an interest in the development of bicycle electric lighting. It is also gratifying to know that, while the prices of earlier mediums continue to rise, good quality and well-manufactured electric-powered lamps from the 1920s and 1930s can still be collected and enjoyed without the need to apply for a bank overdraft. Happy collecting!.

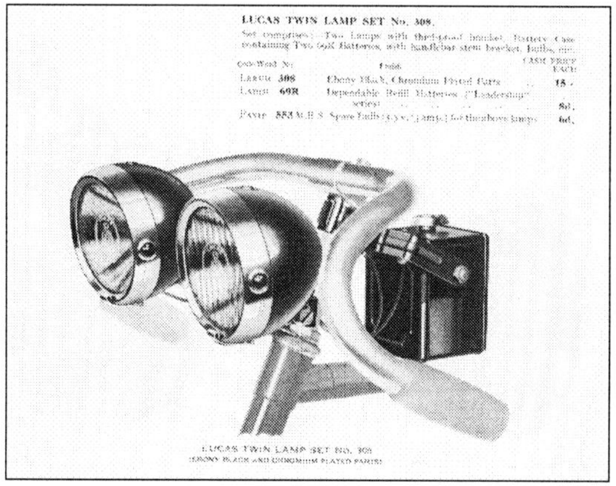

References:

References:

Edison Swan Electric Company Ltd. The Pageant of the Lamp, Thomas Waide – Leeds, 1946.

Patent Office Centenary. A story of the first 100 years in the life of the patent office, H. Harding.

The Cycle Trade Journal 1892-1900

Cycling Magazine 1898-1939

The 'Accessory' Magazine 1909-1911

Various catalogues, broad-sheets, advertisements and brochures as featured.

*We cannot live in the past, and yet from it
we can draw the strength and the wisdom to face the future.*

CATALOGUES & ADVERTISMENTS

Catalogues and advertisments reproduced in this section are arranged by manufacturer and then in date order:

"The Light that Never Fails"

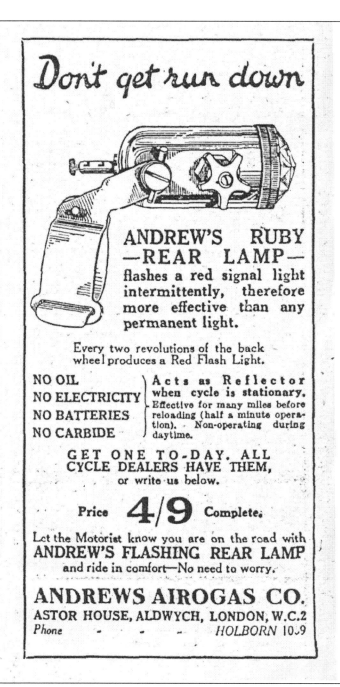

Acme Electric Lamp Co 1898

Andrews Airogas 1927

THE
ARABIAN ELECTRIC CYCLE LAMP.

GIVING A LIGHT FOR FOUR HOURS.

One-Third Actual Size.
Ready for Fixing to Lamp Bracket.

THE ARABIAN ELECTRIC CYCLE LAMP,
Fixed on Lamp Bracket (Giving a Light for 4 Hours with one Charge).

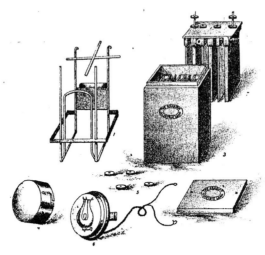

PARTS OF THE LAMP.

1. Bicycle carrier for battery.
2. Inner lid with the elements, zincs and carbons attached (The thick plates are carbons and thin plates zincs.)
3. Battery cells with porous pots in position.
4. Outer lid of battery.
5. Screws for bicycle carrier and terminals.
6. Back of lamp holding incandescent globe, and showing switch.
7. Lens of lamp. (Detached from lamp to show method of fastening.)

Arabian Electric Cycle Lamp Co 1898

Fixing and Wiring Instructions

The fitting and wiring of the "BULLI-MONTIL" Set as indicated on the diagram below is quite a simple matter. By carefully following the instructions the cyclist should quickly complete the task without experiencing any difficulty whatever.

1. Fit Dynamo on left-hand side of fork and adjust so that the small rubber wheel is about ⅛ in. from the RIM of cycle wheel. To obtain correct clearance from spokes and valve, loosen slightly the Band Clip H encircling the Dynamo (see illustration previous page) and move to left or right as necessary. Once the correct position is obtained tighten the Band Clip up again.

2. Fit Head and Rear Lamps on respective brackets, care being taken to pierce the enamel on tubes by the small Earthing Screws G. Connect Dynamo to Head Lamp by SHORT Cable, and Head Lamp to Rear Lamp by LONG Cable.

3. When Dynamo is required in working position it is only necessary to press the plated stud B, which automatically brings the small rubber wheel in correct contact with the rim ready for generating light. When first tested, should no light be generated, it can only be through the small pointed Earthing Screws G not having pierced the enamel on the fork, thus not making a complete circuit.

Retail Price : 35/- Front & Rear, 30/- Front.

(Complete with Head and Rear Lamps, Dynamo, Cables, Cable Bands, 2 spare Bulbs in case, and small Spanner for fixing.)

Obtainable Of All Agents,

also

Wholesale from all the leading Factors, or

The MONTIL MANUFACTURING CO. Limited.

LOWER TRINITY STREET,   BIRMINGHAM.

Telephone: Central 1250.   Telegrams: Pedals, Birmingham.

Also at London, Manchester and Glasgow.

Supplement to the "C.T.C. GAZETTE." September, 1922.

BULLI·MONTIL

CYCLE

Electric Light Set

"Lighting that costs you nothing"

Bulli-Montil 1922 (1)

STARTLING though the statement may at first appear, the "BULLI-MONTIL" Electric Light Set offers "Lighting that costs you nothing."

Differing from any other Cycle Lighting plant, lighting expenses completely end with the initial purchase, with the slight exception of occasional renewal of the small rubber wheel driving the dynamo.

The easy-running, ball-bearing, self-contained dynamo fits ANY standard shape of front fork, requires no extra power to drive, and generates a brilliant, steady and reliable light.

Compared with the "BULLI-MONTIL,"
other Electric Lighting Sets are but "toys."

The "BULLI-MONTIL" sets up a new standard in Cycle Lighting by Electricity, and for consistent reliability there are none to equal it.

It reaches the highest standard of electrical efficiency, and is specially designed and built of finest grade electrical materials with the utmost precision and care throughout. Perfect in every detail, it will outlast the best grade bicycle, and give effective service all the time.

Indestructible — and cannot burn out.

By reason of exclusive design and patent construction, the "BULLI-MONTIL" Dynamo cannot burn out. This is a point of vital importance—and one in which the "BULLI-MONTIL" differs fundamentally from any other set on the market.

In ALL weathers and at ALL speeds it is perfectly safe and reliable. It cannot get out of order, and when once correctly fitted requires no attention whatever beyond an occasional oiling of the dynamo bearing —a distinct advantage over any other form of lighting.

Perfect in Principle and with
nothing to get out of order.
Fully dependable and fool-proof.

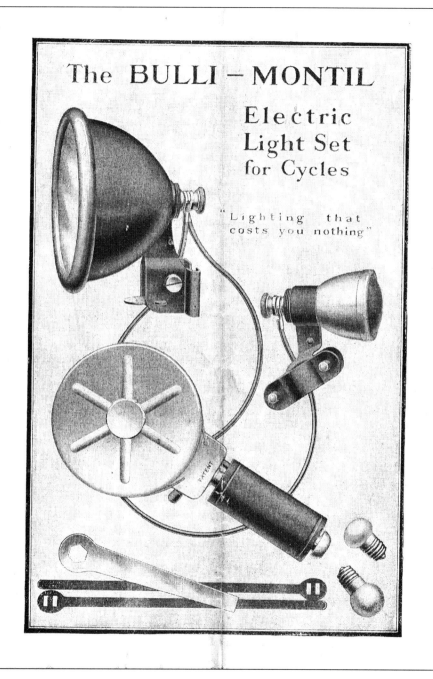

Bulli-Montil 1922 (2)

Special "BULLI-MONTIL" Features.

The following points should be specially noted. They include features found in no other set—features which demonstrate the superiority of the "BULLI-MONTIL" and contribute largely to its remarkable efficiency.

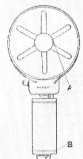

The DYNAMO.

It should be remembered that Dynamo adjustments are NEVER necessary. The Dynamo is a self-contained unit running on ball-bearings and cannot get out of order, and on no consideration must it be tampered with. Beyond the insertion of a few spots of high-class lubricating oil at the ball-oiler A every three months it needs no attention whatever.

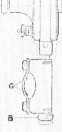

Spring Catch Release.

When fitted (see back page) the Dynamo automatically gets in correct position for driving off the bicycle wheel rim, simply by pressing the plated stud B.

When Lighting is not required, the Dynamo assumes the "free" or "off" position as indicated by the dotted lines in sketch at side.

ADJUSTABLE FOCUS.

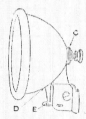

To guarantee perfect focus and maximum lighting the head lamp is fitted with a special adjustable focussing arrangement. The terminal C carrying Bulb is movable horizontally, and correct focus is obtained when, by moving the Bulb backwards or forwards as the case may be, the light coming through the small hole D in lamp body, shines exactly on the red line E on small plate provided for that purpose.

EARTHING SCREWS.

To ensure a complete electrical circuit, Dynamo and Rear Lamp are provided with small Earthing Screws G. When fitting, it is extremely important to make sure that these screws penetrate the enamel on the frame tubes. Failure in this particular will cause trouble, but with a little care perfect and lasting contact is readily assured.

SPECIAL NOTE. The use of a Dry Battery will cause the Dynamo to become De-magnetised, unless used in conjunction with our specially Wired and Designed Battery Case. PRICE 6/-

Bulli-Montil 1922 (3)

The finest Lighting Investment
ever offered to the Cyclist.

Giving an infinitely cleaner, smarter and more reliable method of lighting, and complete immunity from the mess and bother associated with oil or acetylene lamps, the "BULLI-MONTIL" Electric Light Set at the moderate retail price 35/- is a really remarkable proposition.

It ensures perfect lighting fore and aft, and every Cycle owner should consider these

Notable "BULLI-MONTIL" facts.

The only set to solve in a thoroughly practical and dependable manner the problem of efficient and reliable Cycle Lighting by Electricity.

Represents the highest point ever attained in electrical efficiency, and is specially built of the best grade electrical materials.

Handy, compact, dirt, dust, and trouble-proof. Cannot get out of order, and needs no attention. Small rubber wheel on dynamo easily replaced when worn out.

Absolutely self-contained and complete in every way. There are no extras or recurring upkeep charges.

Spring Catch provides immediate engagement with rim of wheel when lighting is required. Only the work of a moment to put "out of action" again.

Requires practically no extra power to drive, and weighs no more than the average acetylene or oil lighting set.

SPECIFICATION.

The DYNAMO. Of exclusive design and manufacture. Self-contained, fully dependable and reliable. Indestructible, and will not burn out. Ball-bearings. Guaranteed 100 per cent. efficiency. Dirt-dust-and-rain-proof, and safe at ALL speeds. Will out-last the machine. Only attention required is an occasional spot of good class oil inserted through the special ball-oiler. Best quality ebony and nickel finish.

The HEAD LAMP. 3½in. dia. fitted with special parabolic silver-plated reflector ensuring a powerful and perfect beam of light. Special adjustable focussing device automatically indicates correct focus for maximum lighting. High-class black enamel, egg-shell finish, and nickel-plated parts.

The REAR LAMP. 1¼in. dia. fitted with best quality solid ruby lens. Black enamel, egg-shell finish, and nickel-plated parts.

CABLES. Specially strong, weatherproof, black finish. Ample size connections.

COMPLETE with small spanner and two spare bulbs—one each for front and rear lamps—of special construction which will not burn out.

SPECIAL NOTE. For Efficiency's sake the
DYNAMO MUST NOT BE TAMPERED WITH.
Cannot be dismantled. NEVER requires attention.

Bulli-Montil 1922 (4)

Fig. 1.—NEW ELECTRIC BICYCLE LAMP.

ECLIPSE ELECTRIC LAMP.

This electric lamp is without doubt the best of its class that has yet been brought out for the use of cyclists. It will throw a brilliant light 75 to 100 feet ahead of the wheel, and the light cannot possibly be jarred or put out unless turned out by the rider. It is as readily attached to a wheel as an oil lamp and no more difficult to operate. The current is provided by a material put up in small cans to fit into a tool bag, and is in the shape of dry powder and known as Eclipse electric sand. It costs no more than oil.

Chloride 1878

Eclipse 1878

Electric Cycle Lamp.

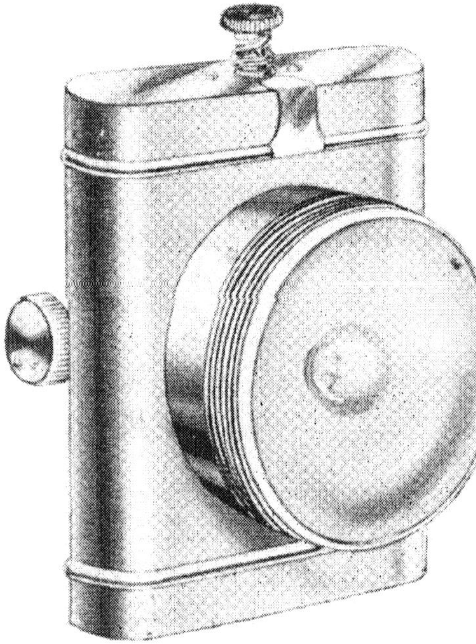

The . .
"EVER READY"
Popular
Model.

Each Battery will give 10 hours of light when used intermittently, and the best results are obtained when the Lamp is used for short periods.

Plain Bevelled Glass, ... Price, 6/6 complete.
If fitted with Bull's-eye Lens, ... ,, 7/6 each.
Spare Batteries, ... ,, 1/- ,,
Spare Metallic Filament Bulbs, ... ,, 1/6 ,,

Constructed throughout of solid brass, heavily nickel-plated. London made. Weight, complete with Battery and Bulb, 11 ozs. Height, 3½ ins.

"Ever Ready" Standard Model Electric Cycle Lamp, each 8/6.

No responsibility is acknowledged in respect to Electric Lamps.

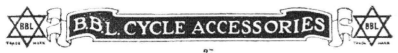

BBL B.B.L. CYCLE ACCESSORIES BBL
TRADE MARK TRADE MARK

27

Ever Ready 1910

EVER READY Electric BICYCLE LAMPS & LANTERNS

Type.	Finish	Reflector Diam.	Dimensions L.	W.	H.	Weight ozs.	Battery Type	Battery No.	Bulb No.	List No.	Price s.	d.
BICYCLE LAMPS	SELF-CONTAINED	MODELS										
1929 Long Life Model	Stoved Black	3"	3"	3½"	4"		2-cell	800	1896	2236	4	3
Spotlight	Stoved Black	2¼"	2¾"	3¼"	3¾"	16	2-cell	800	1428	2036	3	3
Unit Cell Model	Black Enamel	2¼"	3"	3"	3½"	14	do.	U2	2111	2126	4	3
Ever Ready Model	Nickel Plated	2¼"	2¾"	3¼"	4"	18	do.	800	1896	1500	9	6
do. with Bull's-eye	do.	2¼"	2¾"	4¼"	4"	20½	do.	800	1896	1500a	10	6
do. Spotlight	do.	2¼"	2¾"	3¼"	4"	18	do.	800	2111	1500s	10	0
Small E.R. Model	do.	1¾"	2½"	2¾"	3½"	11	do.	1276	1428	1550	7	6
do. with Bull's-eye	do.	1¾"	2½"	3¼"	3½"	12	do.	1276	1428	1523	8	6
Rear Lamp	Stoved Black	1⅜"	4½"	1⅜"	3½"	7½	Unit	U2	1891	2136	2	6
Large 3-cell Model	do.	2¼"	3½"	3¼"	4½"	18	3-cell	295	1908	2035	6	6
MODELS WITH SEPARATE BATTERY CONTAINER												
Front and Rear Set	Leather Bag Nickel Fittings	1⅜"F 1⅜"R	8¾"	1¾"	3"	32	2-cell	1921	R.1428 F.1896	1921	21	0
As above but with	Front Lamp only						2-cell	1921	1896	2020	17	6
Front and Rear Set	Black Leather Nickel Fittings	2¼"F 1⅜"R	4⅜"	2"	4½"	30	F 2-cell R Unit do.	8126 U13	1428 1428	2132	10	6
As above but with	Front Lamp only						3-cell	126	1429	2032	7	6
Bull's-eye Model	Black Leather Nickel Fittings	2¼"	4⅜"	2"	4½"	28	do.	126	1908	2031	12	6
Spotlight Model as	No. 2031						do.	126	2112	2031s	14	0
Bull's-eye Model	Black Leather	1⅜"	4⅜"	2"	4½"		do.	126	1908	2030	12	0
The sets, with separate battery containers, can be supplied with accumulators.												
INSPECTION LAMPS with separate Battery Container.												
Bull's-eye Lens	Black Leather	1⅜"	4⅜"	2"	4½"	20	3-cell	126	1908	1830	9	6
Spotlight Model	do.	1⅜"	4⅜"	2"	4½"	21	do.	126	2112	1830a	10	6
Cap and Lapel Light	do.	1"	4⅜"	2"	4½"	17	do.	126	1908	2042	7	6
Meter Reading Lp. Spotlight Reflector	Stoved Black Nickel Fittings (fitted with Belt Clip)	1½"	3½"	1⅜"	4"	19	do.	295	2112	2090	12	6
Night Shooting Lamp	(as No. 2090, with leather strap on reflector)								2112	2091	15	0
POLICE LAMPS												
Service Model	Stoved Black	2"	3¾"	3½"	6½"	28	3-cell	592	2112	2059	15	0
do.	do. fitted with accumulator						4 volt	2060	2112	2059a	25	6
Standard Model	do.	1¼"	3⅝"	2¼"	6"	20	3-cell	295	1429	1859	8	6
do. Spotlight	do.	1¼"	3⅝"	2¼"	6"	20	do.	295	2112	1859sl	10	0
do. do.	No. 1859sl fitted with Ruby Screen						do.	295	2112	2159	12	6
LANTERNS (Self-contained)												
Cycle Lamp Model	Leather Covered	2¼"	2½"	1½"	3⅝"	17	2-cell	800	1896	878	10	6
do. Prismatic	Reflector	2¼"	2⅝"	1½"	3⅝"	17	do.	800	1896	878a	12	6
Large Handle Type	Nickel Plated	2¼"		3"	3⅞"	28	3-cell	1757	1899	1820	12	6
do. Small Model	do.	1¼"		1⅞"	2⅞"	10	do.	1755	1907	1920	9	6
UprightReflectorTyp.	do.		2" diam.		7¾"	11	do.	1755	1898	1755	10	6
do.	Leather Covered		2"	,,	7¾"	11	do.	1755	1898	1756	12	0
do. Large Model	Nickel Plated		3"	,,	8½"	27	do.	1757	1899	1757	17	6

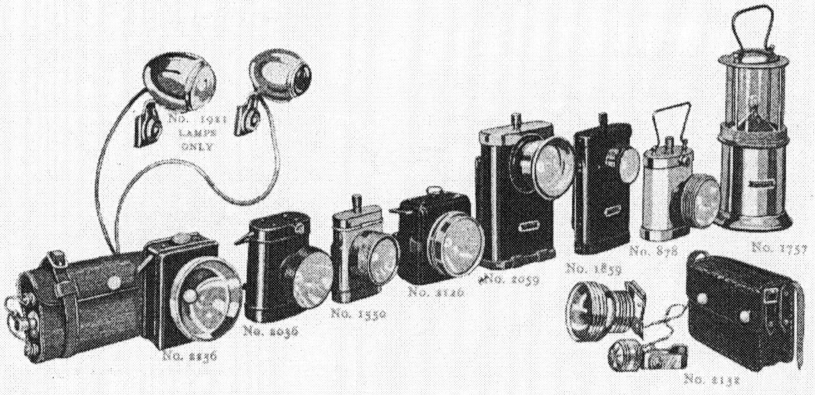

Ever Ready 1930

ELECTRIC CYCLE LAMP.

No. 2036.

To light the lamp, screw down the switch by turning in a clock-wise direction.

To take out a used battery, spring the handle off and remove the top cap. Ever Ready refill No. 800 should be used with this lamp, and should be inserted with the centre brass strip in contact with the bulb.

To fit a new bulb, remove front rim and reflector, then unscrew locking ring behind the bulb. The bulb and locking ring can then be easily removed together. A new bulb, Ever Ready No. 1428, should first be screwed into the locking ring and then both inserted into the lamp together.

Focussing should be effected by screwing the bulb in or out, test with the reflector in position to obtain the desired effect. When the correct position is obtained, the locking ring should be tightened.

When the lamp is in use on a cycle, the handle should be clipped behind the bracket spring to avoid rattle.

THE EVER READY CO. (Great Britain), LTD.

Ever Ready 1936

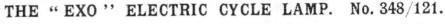

THE "EXO" ELECTRIC CYCLE LAMP. No. 348/121.

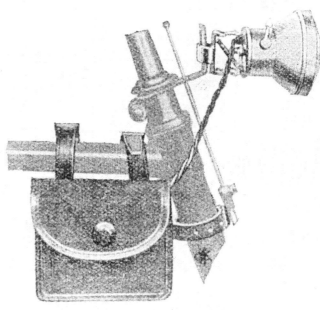

Beautifully nickel plated. Fitted with bulls eye front glass. Fitted with canvas case for carrying battery, best metal filament Bulb. Flexible wire. Best 4 volt, 2 ampere, Accumulator which can be charged at any Electric light station when run down.

Burns 2 hours continuously with one charge.

Cost of re-charging is 4d.

After Accumulator has been charged two or three times it will give 4 hours' continuous light.

PRICE :

Complete and ready charged for immediate use, **8/- each.**

We recommend customers to have a spare accumulator as they can then use the same when other is being charged.

Price 2/6 each.

We will recharge Accumulators for 6d. each.

Exo Electric Cycle Lamp 1912

Lampe Électrique "AUTOWATT"

NOUVEAUTÉ

SENSATIONNELLE

D'une construction particulièrement soignée, c'est le seul appareil donnant en toute sécurité et sans aucune peine, un éclairage électrique.

La Magnéto

AUTOWATT

spécialement étudiée en vue d'une longue durée et d'un fonctionnement parfait, réalise enfin le rêve de ceux que les anciennes lanternes à huile, à pétrole et à acétylène ont si souvent tourmentés.

Grâce à sa simplicité, à ses roulements à billes réglables et à son carter étanche, elle fonctionne également bien par tous temps et ne nécessite aucun entretien.

L'appareil complet en ordre de marche comprenant la magnéto, son support, la lanterne avec sa lampe et les câbles de connexion, en aluminium poli, au prix exceptionnel de **19.50**

Émaillé couleur........ **22.50**

French Brochire for1914

LANTERNES ÉLECTRIQUES

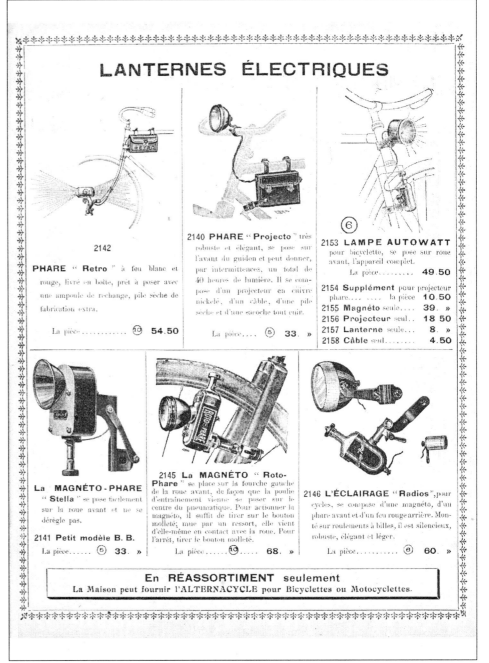

2142

PHARE "Retro" à feu blanc et rouge, livré en boîte, prêt à poser avec une ampoule de rechange, pile sèche de fabrication extra.

La pièce .... ⑩ **54.50**

2140 PHARE "Projecto" très robuste et élégant, se pose sur l'avant du guidon et peut donner, par intermittences, un total de 40 heures de lumière. Il se compose d'un projecteur en cuivre nickelé, d'un câble, d'une pile sèche et d'une sacoche tout cuir.

La pièce .... ⑤ **33.** »

⑥ **2153 LAMPE AUTOWATT** pour bicyclette, se pose sur roue avant, l'appareil complet.

La pièce........ **49.50**

2154 Supplément pour projecteur phare.... .... la pièce **10.50**
2155 Magnéto seule.... **39.** »
2156 Projecteur seul.. **18.50**
2157 Lanterne seule.... **8.** »
2158 Câble seul........ **4.50**

La MAGNÉTO-PHARE "Stella" se pose facilement sur la roue avant et ne se dérègle pas.

2141 Petit modèle B. B.
La pièce...... ⑤ **33.** »

2145 La MAGNÉTO "Roto-Phare" se place sur la fourche gauche de la roue avant, de façon que la poulie d'entrainement vienne se poser sur le centre du pneumatique. Pour actionner la magnéto, il suffit de tirer sur le bouton molleté; mue par un ressort, elle vient d'elle-même en contact avec la roue. Pour l'arrêt, tirer le bouton molleté.

La pièce...... ⑩ .... **68.** »

2146 L'ÉCLAIRAGE "Radios", pour cycles, se compose d'une magnéto, d'un phare avant et d'un feu rouge arrière. Monté sur roulements à billes, il est silencieux, robuste, élégant et léger.

La pièce........ ⑧ .... **60.** »

| En RÉASSORTIMENT seulement |
| La Maison peut fournir l'ALTERNACYCLE pour Bicyclettes ou Motocyclettes. |

French Catalogue for1925 (1)

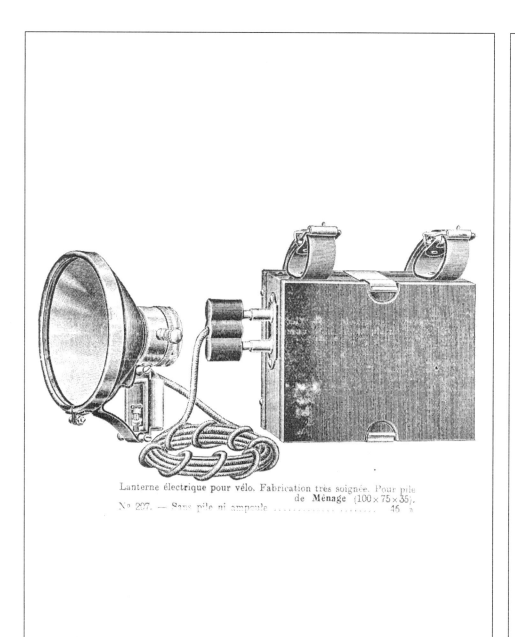

ELECTRIC CYCLE LAMPS.

V 2036 Ever Ready.
Price **3/6** each.
With Battery and Bulb.
Case only, **2/4**

V 69 Lucas.
Price **3/6** each.
With Battery and Bulb.
Case only, **2/4**

V 30 P. & H.
Price **3/6** each.
With Battery and Bulb.
Case only, **2/4**

SPARE BATTERIES.	SPARE BULBS.
V 3819 Ever Ready, 2½ volts, **1/-** each.	**V 4450** Opal Bulb ... **2d.** each.
	V 8473 Phillips' Helix ... **10d.** ,,

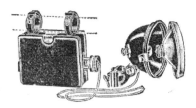

V 8896 A great favourite. Long Life Battery is contained in a metal case which has straps to fasten on to the frame of a cycle or motor cycle. Removable Bulb Holder and extra strong Screw Bracket.

Price, complete with Battery and " Whitelite " Bulb **4/6**

Less Battery and Bulb **3/1½**

V 8097 Cheaper quality, with Spring Bracket and Fixed Reflector.

Price, complete with Battery and V 3568 Bulb **3/11**

Less Battery and Bulb **2/9**

SPARE BATTERIES.	SPARE BULBS.
V 6840 Halford Long Life ... **1/-**	**V 3568** Opal, 3.5 volts ... ... **2d.**
	V 9101 " Whitelite," 3.5 volts ...**4½d.**

HALFORD'S for Service.

French Catalogue for1925 (2)

Halford Cycle Company Ltd 1929 (1)

Lanterne électrique pour vélo. Fabrication très soignée. Pour pile de **Ménage** (100 × 75 × 35).

Nº 207. — Sans pile ni ampoule .... 45 »

56   THE HALFORD CYCLE CO., LTD.

ELECTRIC CYCLE LAMPS (Front)

Ideal for short runs or for cycling to and from work. The small batteries supplied with these Lamps are not made to give continuous light over long periods.

V 8471 A popular side-opening Lamp. Neat front switch. Price, complete with Battery V 6272, and Bulb V 3568 ... ... **1/10**
Price, less Battery and Bulb ... **1/3**

V 8470 New Pattern Lamp, 2 Standard Batteries in parallel, giving longer burning hours.
Price, with V 6272 Batteries and V 3568 Bulb **3/-**
Price, less Batteries and Bulb **2/-**

ELECTRIC TAIL LAMPS

V 9108 Neat pattern, Rear Stay fitting **1/-** each.
V 9108C Complete with Battery, Bulb and metal case. ... ... **3/6** set.

V 2136 Ever Ready, with reliable unit cell. Price **2/6** complete.
Less Battery and Bulb ... **1/9**
Spare Battery VU 2, 1½ volts, **4½d.**
Spare Bulb V 1891, 2.5 volts, **4½d.**

V 9109 Metal case for **Halford** long life Batteries V 6840 and cord ... **1/9** each. When fitting this case to cycle frame, see the small screws (on clips) penetrate the enamel and thus make an earth connection.

Halford Cycle Company Ltd 1929 (2)

The "Helco" Dynamo
(Combined Dynamo and Rear Lamp)
Cycle Lighting Set

(Prov. Patent No. 12144/22).

Solves the Rear Light Question.

The "Helco" Dynamo Lighting Set is designed and manufactured by engineers, is beautifully made, and can be fitted to any cycle in five minutes.

The dynamo and rear lamp are enclosed in an aluminium die casting, highly polished. The front lamp is brass, heavily plated.

The Set lights at **SLOW** walking pace, and **CANNOT BURN OUT AT ANY SPEED.**

We will replace, free of charge, any of our dynamos that prove unsatisfactory through faulty materials or workmanship.

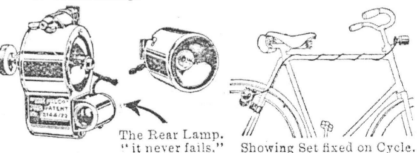

The Rear Lamp. "it never fails."   Showing Set fixed on Cycle.

Don't worry about your rear light! Use the "Helco," and **know positively** that if the head lamp is all right, the rear lamp **MUST BE!**

18/6
Postage 9d.

Stocked by all the leading Cycle Agents, or direct from the manufacturers—
HELLIWELL & CO.,
349, Bristol Rd., Birmingham.

Helco Dynamo advertisement for 1922

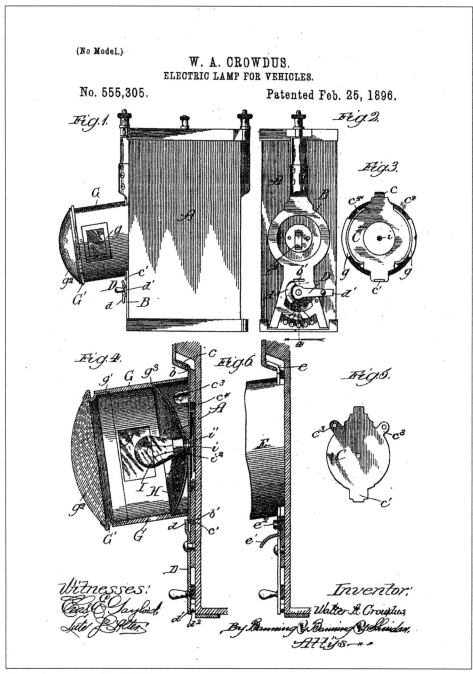

(No Model.)

W. A. CROWDUS.
ELECTRIC LAMP FOR VEHICLES.

No. 555,305.   Patented Feb. 25, 1896.

Fig.1.   Fig.2.   Fig.3.

Fig.4.   Fig.6.   Fig.5.

Witnesses: Inventor:
                                 Walter A. Crowdus
                                 By Banning & Banning & Sheridan,
                                 Att'ys

(No Model.)   2 Sheets—Sheet 1.

B. B. HOFFMAN.
ELECTRIC LIGHT FOR BICYCLES.

No. 559,801.   Patented May 12, 1896.

Crowdus patent for 1896

Hoffman patent for 1896

33

CATALOGUES & ADVERTISMENTS

LAMPS — FRONT and REAR

EVER READY No. 2308

Rust-proofed all-metal stove-enamelled case, in black or silver, octagonal ribbed ruby plastic lens and positive indicator type switch.

A1210/302 Black... Each 4/9
A1211/302 Silver... „ 4/9

EVER READY No. 2307
Front with fork bracket.
A1220/302 Black... Each 4/9
A1221/302 Silver... „ 4/9

STARLITE FRONT LAMP No. 1085

A solidly made, well finished lamp, at a reasonable price.
A1229/3206 Silver finish ... Dozen 48/9

STARLITE FRONT LAMP No. 1289

Strongly made with switch at top.
A1230/3508 Silver ... Dozen 53/6

EVER READY No. 2306

All-metal stove-enamelled rust-proofed case, in range of colours. Positive indicator type switch.
A1212/210 Black ... Each 4/3
A1213/210 Silver ... „ 4/3
A1214/210 Blue ... „ 4/3
A1215/210 Green ... „ 4/3

EVER READY No. 2339

Extruded aluminium case, in range of colours. Bottom cap switch and loading. Lens 1⅜" diameter.
A1222/110 Black ... Each 2/9
A1223/110 Silver ... „ 2/9
A1224/110 Blue ... „ 2/9
A1225/110 Green ... „ 2/9

EVER READY No. 2236

Black or silver stove-enamelled all-metal case, rust-proofed. Positive indicator type switch and 3" diameter reflector with chrome snap-on front.
A1227/506 Black ... Each 8/3
A1228/506 Silver ... „ 8/3

EVER READY No. 2536

All-metal stove-enamelled case, rust-proofed. Positive indicator type switch, and octagonal plastic lens.
A1216/300 Black ... Each 4/6
A1217/300 Silver ... „ 4/6

No. 2536 Screw Switch
A1218/300 Black ... Each 4/6
A1219/300 Silver ... „ 4/6

STARLITE COMBINATION SET No. 1211

Battery combination set, comprising front and rear lamp with contact and terminal attachment.
A1231/6104 Silver finish ... Dozen 92/-

STARLITE REAR LAMP No. 1280

Heavy duty rear lamp with switch at top, as preferred by many users.
A1232/3200 Silver finish ... Dozen 48/-

Leicester Liverpool Luton Manchester Middlesbrough Morecambe Oxford Perivale Sheffield Stratford, E.15 77

Kerry's of GB Ltd (post) 1948 (1)

LAMPS — FRONT and REAR. DYNOHUBS

PHILLIPS MODEL TE511 FRONT LAMP

With sliding switch and 2½" reflector.
A1240/210 Black ... Each 4/3
A1241/210 Silver ... „ 4/3

MODEL TE512 FRONT LAMP

Similar to model 511, but with screw-down switch.
A1242/210 Black ... Each 4/3
A1243/210 Silver ... „ 4/3

D.B.U.

ROADSTER HEAD LAMP

A1249/1108 Chromium plated Each 17/6

FOR COMPLETE WHEELS SEE PAGE 128

PHILLIPS REAR LAMP No. TE611

Superbly finished in lustrous silver enamel. Screw-down switch.
A1244/208 Silver ... Each 4/-

STURMEY ARCHER

G.H.6 6V "DYNOHUB" LIGHTING UNIT

6-volt Dynohub designed to be mechanically free of friction, ensuring trouble-free running.
A1250/4105 GH6 ... Each 57/1
A1251/605 DBU extra ... „ 8/9

GEAR AND DYNOHUB

Combined 6-volt Dyno and 3- or 4-speed wide-ratio hub in one unit.
A1252/6707 AG 3-speed ... Each 93/2
A1253/7209 FG 4-speed ... „ 100/4
A1254/605 DBU extra ... „ 8/9

PHILLIPS MODEL TE521 FRONT LAMP

With sliding switch and 3" reflector.
A1245/406 Black ... Each 6/9
A1246/406 Silver ... „ 6/9

MODEL TE522 FRONT LAMP

Similar to model 521 but with screw-down switch.
A1247/406 Black ... Each 6/9
A1248/406 Silver ... „ 6/9

SPORTS HEAD LAMP
Light-weight streamlined model, chromium plated.
A1255/1108 ... Each 17/6

REAR LAMP

Improved design, now fully streamlined.
A1256/206 M6 R Roadster (oval) Each 3/9
A1257/206 M6 S Sports (round) Each 3/9

78 Birmingham Blackburn Brighton Cambridge Canterbury Chester Colwyn Bay Croydon Exeter Ipswich

Kerry's of GB Ltd (post) 1948 (2)

34

LAMPS — DYNAMO SETS

LUCAS

"KING MINOR"

A small light-weight model with 2¾" diameter moulded light unit giving a powerful spot beam. Supplied complete with twin cable and No. CD.36 dynamo with combined CT.88 tail lamp.

A1275/2500 Chromium plated 37/6
A1276/2304 Silver lustre ... 35/-

Head Lamp HC.5

A1277/1000 Chromium plated 15/-
A1278/804 Silver lustre ... 12/6

HEAD and TAIL LAMP SET No. 431

For mopeds and motor-assisted cycles. For use with engines which incorporate a lighting coil in the flywheel generator. Supplied complete with specially matched bulbs, twin cables with connecting nipples and terminals.

A1279/1704 Black with chromium plated rim 26/-
A1280/1704 Silver with chromium plated rim 26/-

REAR LAMP No. CT 88

Streamlined model with grey plastic body and large-diameter lens, conforming to the new rear lighting regulations. Lens is detachable for bulb replacement.

A1281/304 ... ... 5/-
A1282/06 Separate lenses 9d.

THE NEW "PATHFINDER"

A new cycle dynamo set with chromium plated hinged front and hooded rim. Provision in the body for a stand-by battery operated by "push-push" switch. Complete with twin cable and No. CD.36 dynamo with combined No. CT.88 tail lamp.

A1283/3304 Chromium plated ... 50/-
A1284/3000 Silver lustre ... 45/-

Head Lamp HC.11

A1285/1704 Chromium plated ... 26/-
A1286/1400 Silver lustre ... 21/-

THE NEW "CAPTAIN"

Attractive chromium plated torpedo-shaped head lamp of completely new design. Set includes dynamo and re-designed rear lamp.

A1287/2500 Chromium plated ... 37/6

Head Lamp HC.13

A1288/1000 Chromium plated ... 15/-

Rear Lamp CT98

A1288a/400 ... ... 6/-

SET No. 683

Still a firm favourite, this set includes the No. 596 head lamp, twin cable and No. CD.36 dynamo with combined No. CT.88 tail lamp.

A1289/2500 Chromium plated ... 37/6
A1290/2304 Silver lustre with chromium plated rim ... 35/-

"KING SPORTS"

Streamlined light-weight model appealing to sports riders. Moulding of lens together with silver-plated reflector gives an intense spot beam. Set supplied complete with twin cable and No. CD.36 dynamo with combined CT.88 tail lamp.

A1291/2608 Chromium plated ... 40/-
A1292/2500 Silver lustre ... 37/6

Head Lamp HC.4

A1293/1108 Chromium plated ... 17/6
A1294/1000 Silver lustre ... 15/-

HEAD and TAIL LAMP SET No. 331

For mopeds and power-assisted cycles. Comprises No. 313 head lamp with provision for stand-by battery and VT.31 tail lamp. For use with engine units which incorporate a lighting coil and flywheel generator.

A1295/1704 Silver lustre with chromium plated rim 26/-

VT31 REAR LAMP

A1296/304 Silver 5/-

"MINILAMP"

Neat and attractive lamp of aerodynamic design. Moulded independent lens, strong plastic body. Colours, blue, yellow, grey, red, green, black, white.

A1297/304 Less bulb 5/-
Bracket for fitting under handlebar expander bolt.
A1298/04 ... ... 6d.

LAMPS — DYNAMO SETS

MILLER

DYNAMO SET No. 536T

Finished in silver grey or lustrous black enamel with chrome rim, with provision for stand-by battery.

A1260/2608 536T. Black ... Each 40/-
A1261/2608 536T. Silver ... „ 40/-

HEAD LAMPS

A1262/1104 No. 6 Silver ... Each 17/-
A1263/1104 No. 6 Black ... „ 17/-

DYNAMO SET No. 537T

Similar to model 536T described above, but heavily chromium plated.

A1264/3000 ... ... Each 45/-

HEAD LAMP

No. 7 Chromium plated.

A1265/1408 ... ... Each 22/-

No. 596 REAR LAMP

With panel for number plate illumination as required for motorised cycles and Mopeds. Chromium plated.

A1266/308 ... ... Each 5/6

No. 599 REAR LAMP

Chromium plated, clip fitting.

A1268/304 ... ... Each 5/-

DYNAMO SET No. 535T

Weighing only 5¼ ounces, this ultra-light-weight head lamp is heavily chromium plated. The dynamo is combined with the tail lamp for rear wheel fitting.

A1269/2304 ... ... Each 35/-

No. 5 HEAD LAMP

Chromium plated.

A1270/800 ... ... each 12/-

SET No. 6TM

For motorised bicycles. Comprising the No. 6 Miller head lamp with provision for stand-by battery, and No. 596T tail lamp with panel, for number plate illumination. Boxed complete with bulbs and cable.

A1267/1608 Black or Silver ... Each 25/-

DYNAMO SET No. 539T

A reasonably priced cycle lighting set. Torpedo-shaped head lamp, finished throughout in chromium plate. The light-weight head lamp is also finished in chromium plate on brass.

Chromium plated.

A1271/2500 ... ... Each 37/6

No. 9 HEAD LAMP

Chromium plated.

A1272/908 ... ... Each 14/6

FOR
LAMP BATTERIES
SEE PAGE 84

FOR
DYNAMO and LAMP
BULBS SEE Pages 85 and 86

Kerry's of GB Ltd (post) 1948 (3)

Kerry's of GB Ltd (post) 1948 (4)

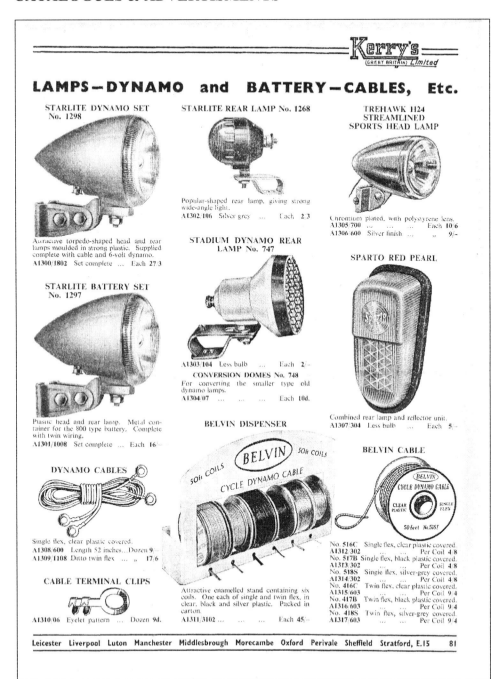

Kerry's of GB Ltd (post) 1948 (5)

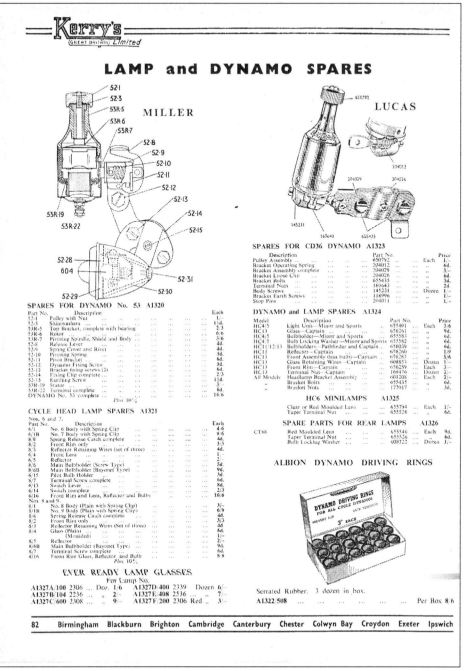

Kerry's of GB Ltd (post) 1948 (6)

Joseph Lucas Ltd Handbill for 1929

We make Light of our Labour

KING OF THE ROAD

LUCAS
"KING OF THE ROAD"
CYCLE . .
DYNAMO
SET Nº 25

The new Lucas Cycle Dynamo Set No. 25 is undoubtedly the finest light-giver for pedal cycles, and is designed to meet the requirements of the cyclist who wishes to have the very best electrical set for his cycle.

Our experience as the largest British manufacturers of electrical equipment for car and motor-cycles is embodied in this Set. It possesses the same reliable characteristics as our car lighting equipment, and is built to a standard and not to a price.

Many years of constant experimenting and research in the Lucas Research Laboratories have gone to build up the Lucas standard of quality.

The Set provides an even, brilliant field of illumination at all speeds, projects a powerful light, even at walking pace, and is an ideal set for touring purposes.

THE COMPLETE SET comprises :—
Dynamo No. 25.
Dynamo Carrier No. DC25.
Head Lamp No. H25.
Tail Lamp No. T25.
Cable and Bulbs.

Code Word.	Finish.			CASH PRICE
LEBEL	Ebony Black, Plated Parts	..	..	£1 5 0
	Spare Bulb B.A.S. No. SS for Head Lamp			1 6
	Spare M.E.S. Bulb for Tail Lamp (3-5 volts)			6

TAIL LAMP
NO T25

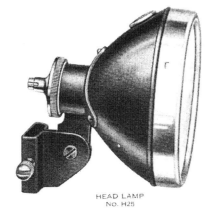

HEAD LAMP
NO. H25

4

We make Light of our Labour

KING OF THE ROAD

Continued.

The **Head Lamp**, of particularly bold and pleasing design, possesses all the refinements of a Lucas Car Head Lamp. It is fitted with a parabolic silvered reflector, and the bulb holder allows the lamp to be correctly focussed. The " Difusa " glass broadens the beam and eliminates all streaks of light.

The **Tail Lamp** for this set is neat in design, gives the usual red light to the rear, and is fitted with our special spring terminal.

The **Dynamo** is of robust construction, is waterproof and dust-proof, and engages with the side of the tire. The clip and spring are ingenious, and it is a simple matter to throw the dynamo into or out of engagement. There are no brushes or commutators to clean, the only attention needed being an occasional lubrica-tion of the bearings. The output of the dynamo is arranged so that there is no overloading of the bulbs under any conditions, thus ensuring that they will give long life.

LUCAS
"KING OF THE ROAD"
CYCLE . .
DYNAMO
SET Nº 25

SET, WITHOUT TAIL LAMP.

Code Word	Finish			CASH PRICE
LEBID	Ebony Black, Plated Parts	..	..	£1 2 6
Dynamo only.				
LEBOT	No. 25, with Carrier Bracket and Cable			16 6
Head Lamp only.				
LECYN	No. H25, with Bulb	..	.. ..	6 0
Tail Lamp only.				
LEDAN	No. T25, with Bulb	..	.. ..	2 6

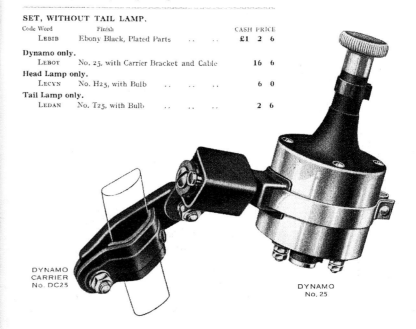

DYNAMO
CARRIER
No. DC25

DYNAMO
No. 25

5

Joseph Lucas Ltd 1927 (1)

Joseph Lucas Ltd 1927 (2)

The Memory of Quality remains

LUCAS
"KING OF THE ROAD"
CYCLE..
DYNAMO
SET Nº 304

SIX VOLT

" THE FINEST LIGHT-GIVER
FOR PEDAL CYCLES "

The Lucas "King of the Road" Cycle Dynamo Sets for pedal cycles are carefully designed in every detail. We have used our extensive experience with Car and Motor-Cycle Lamps and produced, we believe, the most efficient Pedal Cycle Sets that are made. They are sold complete with Dynamo mounting, and can be fitted by any cyclist in a few minutes. They are equally suitable for Tricycles or hand propelled chairs.

They provide a wide field of brilliant and constant light, and make night riding a pleasure. They are always ready the instant they are wanted without any preparation. Ideal for touring, handy for odd night trips, and invaluable for the urgent demands of professional people. Apart from the satisfaction in owning and riding an electrically equipped Cycle, they offer endless light for the very low figure which is their first and final cost.

A special feature of the lamps supplied with these Sets (Nos. 304 and 25) is their large parabolic reflectors of correct optical design, which are covered with a transparent protective coating which prevents tarnishing, and their green side glasses. The bulb holder in each lamp allows it to be accurately focused.

No. 304D Cycle Head Lamp is of bold design, possessing similar refinements to our Car Projectors. A special feature of this lamp is that a dash lamp type of battery is neatly housed behind the reflector. The controlling switch has three positions, as follows :—

ON	Head lamp and tail lamp lit by dynamo.
BATTERY	Head lamp only lit by battery.
DIM	A special dimming device operates when switch is turned to Dim position. The main bulb gives a dimmed light, and tail lamp remains alight.

This dimming device enables the rider to dim his driving light when meeting oncoming traffic, thereby showing courtesy to other road users.

The battery is switched on when the dynamo is not generating, such as when putting the cycle away at night, or when it is wheeled through narrow passages, etc., and it is useful as a hand lamp for roadside repair at night.

The bulbs fitted to this lamp are B.A.S. No. 8DS (6 volt, 0.5 amp.) for the main bulb, and No. 13D M.E.S. (4 volt, 0.3 amp.) for the secondary bulb used with the dry battery.

No. 307 Cycle Head Lamp is distinctive in design and pleasing in contour. The front is held by our patented spring locating clip, and is easily removable for cleaning purposes. The socket fixing is provided with a controlled tightening screw which enables the head lamp to be set at any angle. The bulb in the lamp is a B.A.S. 8S (6 volt, 0.5 amp.). This lamp does not house a stand by battery. (See page 5 for illustration of Lamp.)

No. C25 Dynamo is of improved design. This neat and compact machine is totally enclosed; special precautions are taken to exclude dust and mud from the bearings. The rotor is now mounted on ball bearings which are packed with grease during manufacture and require no lubrication in service. There are no brushes or contactors—simply one thumb screw.

COMPLETE SET No. 304 comprises :—

Dynamo No. C25, with Bracket, Head Lamp No. 304D, Tail Lamp No. CT31, Cables, Bulbs and Battery.

Code Word	No.	Finish	CASH PRICE
LEDUG	304	Ebony Black	£1 3 6
LEDYR	304	Set as above, less Tail Lamp	1 2 0
LEDOL	305R	Dry Battery for above .. .. each	5

HEAD LAMP No. 304D
USED IN No. 304 SET

TAIL LAMP No. CT31
USED IN No. 304 AND
No. 25 SETS

JOSEPH LUCAS LTD., BIRMINGHAM, 19, ENGLAND

4

---

long after Price is forgotten.

continued.

terminal for the Head Lamp and another for the Tail Lamp. The output, as fixed by the design, is constant over a very wide speed range. The Dynamo provides an even, brilliant field of illumination at all speeds and projects a powerful light, even at walking pace.

The Dynamo is usually mounted on the **left hand rear back stay**. Our Spring Tension Spindle Bracket, the spring of which is entirely enclosed, in conjunction with a pen and ratch device, holds the Dynamo out of use. When lights are required it is only necessary to press the release lever towards the Dynamo, when the Dynamo comes into action automatically. A steady quiet drive is ensured by the wide knurled face of the driving pulley being kept in contact with the side of the tyre.

No. CT31 Tail Lamp is very neat and shows an effective red light to the rear, and has adjustable clip fixing for rear stay. It is important that only the standardized bulb be used.

COMPLETE SET No. 25 comprises :—

Dynamo No. C25 with Bracket, Head Lamp No. 307, Tail Lamp No. CT31, Cable and Bulbs.

Code Word	No.	Finish	CASH PRICE
LEBEL	25	Ebony Black, Plated Parts	£1 2 6
LEBIA	25	Set as above, less Tail Lamp	1 1 0
Dynamo only.			
LEBU		No. C25, with Carrier Bracket and Cable	15 0
Head Lamps only.			
Code Word			
LADCE		No. 304D, with Main and Secondary Bulbs	7 0
LEESY		No. 307, with Bulb	6 0
FAECE		Spare Bulb B.A.S. No. 8DS (Diffused) Main Bulb for Head Lamps No. 304D and No. 307	1 4
FAUBU		Spare Bulb No. 13D M.E.S. (4 volt, 0.3 amp.) Secondary Bulb in No. 304D Head Lamps	6
Tail Lamp only.			
Code Word			
LABDA		No. CT31, with Bulb	1 6
FABUE		Spare Bulb No. 3515 M.E.S. (3.5 volts, 0.15 amp.) for Tail Lamp No. CT31	6

LUCAS
"KING OF THE ROAD"
CYCLE..
DYNAMO
SET Nº 25

SIX VOLT

" MAKE NIGHT RIDING
A PLEASURE "

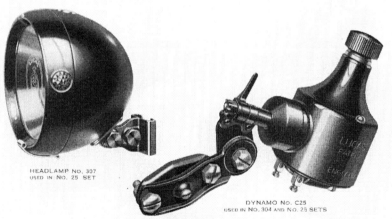

HEADLAMP No. 307
USED IN No. 25 SET

DYNAMO No. C25
USED IN No. 304 AND No. 25 SETS

JOSEPH LUCAS LTD., BIRMINGHAM, 19, ENGLAND

5

Joseph Lucas Ltd 1934 (1)   *Joseph Lucas Ltd 1934 (2)*

King of the Road

LUCAS
"KING OF THE ROAD"
CYCLE . DYNAMO SET Nº 302

FOUR VOLT

Lucas No. 302 Cycle Dynamo Set incorporates our lightweight Cycle Dynamo Type C175. This dynamo possesses all the refinements of the larger types, but its weight has been cut down to a minimum. It will supply current for the head lamp, which is fitted with a 4-volt 0.3 amp. bulb, and also a tail lamp, if required, which has a 4-volt .04 amp. bulb. This dynamo is also fitted with ball bearings.

No. 302 Cycle Head Lamp used in this set has similar characteristics to our No. 304D, but is not fitted with a secondary bulb. It has a large parabolic reflector of correct optical design, heavily silver-plated and non-rusting.

No. CT31 Tail Lamp is very neat, and gives an effective red light to the rear, and has an adjustable clip fixing for the rear stay.

COMPLETE SET No. 302 comprises :—
Dynamo C175 with Bracket, Head Lamp No. 302, Tail Lamp CT31, Cables, Bulbs and Battery.

Code Word	No.	Finish	CASH PRICE
LAEGO	302	Ebony Black	£1 0 0
		As above, **less Tail Lamp**	18 6
Dynamo only.			
LAEPP	C175	With Bracket, Release Lever and Cable	12 0
Head Lamp only.			
LAERK	302	With Bulb	6 6
FATES	43D M.E.S.	Spare Bulb (4 volt, 0.3 amp.) for Head Lamp No. 302	6d.
Tail Lamp, etc.			
LAEDA	CT31	With Bulb	1/6
LEDOL	305R	Dry Battery for No. 302 Lamp	5d.
FATIK	404 M.E.S.	Spare Bulb (4 volt, .04 amp.) for above Tail Lamp when used with No. C175 Dynamo	8d.

TAIL LAMP CT31

HEAD LAMP No. 302

DYNAMO No. C175

JOSEPH LUCAS LTD., BIRMINGHAM, 19, ENGLAND

6

Joseph Lucas Ltd 1934 (3)

King of the Road

Lucas No. 307/4 Cycle Dynamo Set also incorporates our lightweight Cycle Dynamo Type C175, which is a neat and compact machine. Special precautions are taken to exclude dust and mud from the bearings. It is fitted with our Spring Tension Spindle Bracket, similar to the 6-volt Cycle Dynamo No. C25, the spring of which is entirely enclosed. When lights are required it is only necessary to press the release lever towards the dynamo, when the dynamo comes into action automatically.

No. 307 Cycle Head Lamp is distinctive and pleasing in design. It is now fitted with green side glasses, and the front is held by our patent locating spring clip which ensures that the side glasses are always in the correct position. In this set this lamp is fitted with a M.E.S. Bulb No. 43 (4 volt, 0.3 amp.).

No. CT31 Tail Lamp is the same lamp as supplied with No. 302 Set described opposite.

COMPLETE SET No. 307/4 comprises :—
Dynamo C175, with Bracket, Head Lamp No. 307/4, Tail Lamp No. CT31, Cables and Bulbs.

Code Word	No.	Finish	CASH PRICE
LEFEV	307/4	Ebony Black (Plated Parts)	19/-
		As above, **less Tail Lamp**	17/6
Dynamo only.			
LAEPP	C175	With Bracket, Release Lever and Cable..	12/-
Head Lamp only.			
LEFAH	307/4	With Bulb	5/6
Tail Lamp, etc.			
LAEDA	CT31	With Bulb	1/6
FATIR	404 M.E.S.	Bulb (4 volt, .04 amp.) for above Tail Lamp when used with No. C175 Dynamo	8d.

LUCAS
"KING OF THE ROAD"
CYCLE . DYNAMO SET Nº 307/4

FOUR VOLT

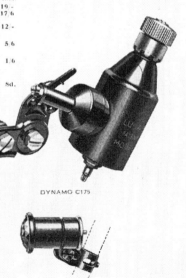

DYNAMO C175

HEAD LAMP No. 307/4

TAIL LAMP No. CT31

JOSEPH LUCAS LTD., BIRMINGHAM, 19, ENGLAND

7

Joseph Lucas Ltd 1934 (4)

LUCAS

"KING OF THE ROAD"
TWIN LAMP
SET No. 308

NEW
DRY BATTERY MODEL

The new Lucas Twin Lamp Set No. 308 is a Dry Battery Lighting Set which brings to cyclists an advantage which hitherto practically only motorists have enjoyed—twin headlamps with a "dim and switch" arrangement.

The twin headlamps are mounted on a horizontal bar with a central slot which fits over the standard handlebar lamp bracket, and incorporates a special thief-proof locking device. Two high-capacity twin cell batteries are housed in a neat cubical battery case carried behind the handlebar in a forked bracket clipped to the handlebar stem. A three-way switch, mounted on the top of the battery case, switches on either one or both lamps, as desired.

This twin lamp set is designed to give much the same effect as the Lucas "Dip and Switch" Reflector System which is fitted to so many cars. The right-hand lamp, fitted with "Difusa" glass, throws a powerful beam ahead, while the left-hand lamp has a fluted glass and, being set at a slight angle, projects a flat beam to the side of the road.

LUCAS TWIN LAMP SET No. 308.

Set comprises :—Two Lamps with thief-proof bracket, Battery Case containing Two 69R Batteries, with handlebar stem bracket, Bulbs, etc.

Code Word	No.	Finish	CASH PRICE EACH
LEBUH	308	Ebony Black, Chromium Plated Parts	15 -
LADIM	69R	Dependable Refill Batteries ("Leadership" series)	8d.
FAVIP	553	M.E.S. Spare Bulb (5.5 v. .3 amp.) for the above lamps	6d.

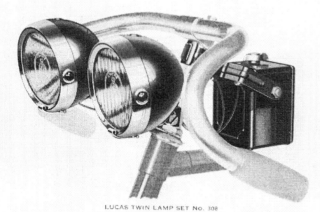

LUCAS TWIN LAMP SET No. 308
(EBONY BLACK AND CHROMIUM PLATED PARTS)

JOSEPH LUCAS LTD., BIRMINGHAM, 19, ENGLAND

8

Joseph Lucas Ltd 1934 (5)

Continued

With both lamps in use, therefore, the cyclist has the advantage of two independent beams, one shining forward and picking out distant objects, the other illuminating the road immediately ahead. When meeting oncoming traffic a turn of the convenient switch dims the near-side lamp and cuts out the right-hand beam. In this way you are able to show courtesy to other road users.

Additionally, of course, the single light can also be used for town riding and other occasions when two lamps are unnecessary, thus economising in current consumption.

In common with practically all the new Lucas Cycle Electric Headlamps these twin lamps are fitted with side glasses and a new patent locating clip for securing the rim. This clip carries a small projecting tongue on its inner side which registers with a hole in the bottom of the rim, preventing it from moving round. With side glasses this is, of course, particularly important.

LUCAS TWIN LAMP SET No. 308.

Set comprises :—Two Lamps with thief-proof bracket, Battery Case containing Two 69R Batteries, with handlebar stem bracket, Bulbs, etc.

Code Word	No.	Finish	CASH PRICE EACH
LERYO	308	Ivory White, Chromium Plated Parts	16.6
LADIM	69R	Dependable Refill Batteries ("Leadership" series)	8d.
FAVIP	553	M.E.S. Spare Bulb (5.5 v. .3 amp.) for the above lamps	6d.

LUCAS

"KING OF THE ROAD"
TWIN LAMP
SET No. 308

NEW
DRY BATTERY MODEL

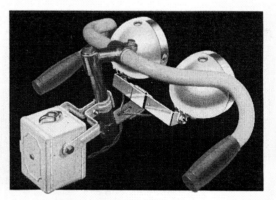

LUCAS TWIN LAMP SET No. 308
(IVORY WHITE, CHROMIUM PLATED PARTS)

JOSEPH LUCAS LTD., BIRMINGHAM, 19, ENGLAND

9

Joseph Lucas Ltd 1934 (6)

LUCAS

"KING OF THE ROAD" BATTERY .. CYCLE LAMP Nº 301 . . .

NEW MODEL

The new Lucas De-Luxe Dry Battery Lamp No. 301 sets a high standard in Dry Battery Lamps. It is imposing in appearance, having a pleasing contour which makes it at once distinctive. A special feature of this lamp is its fluted domed glass which spreads the brilliant beam of light. Its spring back is extremely strong and durable, and absorbs all road shock and eliminates any harm which may be caused to the batteries and bulb when travelling over rough roads. The bulb holder, which is fitted with a 5.5 volt bulb, is hinged, and therefore it is a simple matter to replace the two large-capacity batteries (No. 301R) which are carried on a platform in the body of the lamp, when required. The switch, which is positive in action, enables the rider to dim the bright driving light when cycling in towns or well-lighted roads, and so conserve the batteries' life. It is handsomely finished, all Chromium Plated, and is of that high standard of workmanship and finish always associated with Lucas goods.

LUCAS DRY BATTERY CYCLE LAMP No. 301.

Front, 5 in. diameter.   Aperture, 3¼ in.

Code Word	No.	Finish	CASH PRICE EACH
Lesaf	301	All Chromium Plated, complete with Bulb and Two No. 301R Batteries	12/6
Leset	301R	Dependable Refill Batteries ("Leadership" series)	7d.
Favir	553 M.E.S.	Spare Bulb for above Lamp (5.5 v. .3 amp.)	6d.

LUCAS DRY BATTERY LAMP No. 301

JOSEPH LUCAS LTD., BIRMINGHAM, 19, ENGLAND

10

Joseph Lucas Ltd 1934 (7)

The model **Lucas Battery Cycle Lamp No. 305** is designed to give alternative lights, either a powerful beam for country riding or a reduced light for use when riding in lighted streets, or when the cycle is being wheeled.

The switch controlling these two lights is incorporated in the top of the lamp. It has a very positive action, and it is large enough to be operated easily, even when wearing thick gloves.

The lamp houses two Lucas No. 305R Dry Batteries, both of which are in use in the one switch position, while only one is in use when the reduced light is switched on. This lamp is now fitted with side glasses.

A special feature of this lamp is that a spare bulb is supplied and is carried in a spring clip inside the body of the lamp.

The finish and workmanship of this lamp are of the high quality associated with the name "Lucas." The standard finish is Ebony Black, with the front rim Chromium Plated, giving the lamp a particularly attractive appearance. It can also be supplied in Ivory White finish if ordered.

BATTERY CYCLE LAMP No. 305.

Front, 4 in.   Aperture, 3½ in.   Weight, 22½ ozs.

Code Word	No.	Finish	CASH PRICE EACH
Lesir	305	Ebony Black, Chromium Plated Rim, complete with 305R Batteries and Spare Bulb ready for use	7/6
Lesis	305	Ivory White, Chromium Plated Rim, etc.	8/-
Lebir	305R	Dependable Refill Battery for above Lamp	8d.
Favir	81 M.E.S.	Spare Bulb (8 volts .1 amp.)	6d.

The **Lucas Battery Cycle Lamp No. 310** is an additional model for 1934. Inside the body of the lamp is carried one large-capacity Lucas Dry Battery No. 69R. This model also gives a bright or dimmed light according to the position of the switch. Its square base enables it to be stood upright and used for purposes other than on a cycle. The front glass is fluted, which broadens the width of the beam, and the front rim is fitted with side glasses. A clip is fitted inside the body of the lamp for carrying a spare bulb.

BATTERY CYCLE LAMP No. 310.

Front, 4 in.   Aperture, 3½ in.   Weight, 26 ozs.

Code Word	No.	Finish	CASH PRICE EACH
Lesov	310	Ebony Black, Chromium Plated Rim, complete with No. 69R Battery	7/-
Lesof	310	Ivory White, Chromium Plated Rim, etc.	7/6
Lebix	69R	Dependable Refill Battery for above Lamp	8d.
Favin	282 M.E.S.	Spare Bulb (2.5 volts .2 amps.)	3½d.

LUCAS

BATTERY CYCLE . LAMPS . Nᵒˢ 305 & 310

IMPROVED MODEL AND ADDITIONAL MODEL

LUCAS DRY BATTERY CYCLE LAMP NO. 310

LUCAS DRY BATTERY CYCLE LAMP NO. 305

JOSEPH LUCAS LTD., BIRMINGHAM, 19, ENGLAND

11

Joseph Lucas Ltd 1934 (8)

The Memory of Quality remains

KING OF THE ROAD

LUCAS
BATTERY
CYCLE .
LAMPS.

Nᵒˢ 79 & 69DSB

• ▬▬▬▬ •

WITH SPRING BACKS

This Range of Lucas Dry Battery Lamps, with their exclusive features, create a new and unrivalled standard of excellence. Every lamp body is substantially made of heavy gauge British pressed steel, with a split hinge and overlapping joints, and will, if necessary, stand up to rough treatment. Entirely riveted up, there are no screws to rattle or shake off, and nothing to go wrong.

The New Lucas Dimming Switch is certain in action, easy to manipulate, neat in appearance, and conserves the life of the batteries.

Strong Spring Backs. With the exception of Models Nos. 69D and 69, which have a strong spring-grip socket, the other models have spring loaded backs which eliminate the harm which may be caused to the battery or bulb over a stretch of bad road.

All Lucas Reflectors are accurately designed, heavily Silver-plated and non-rusting.

Chromium-plated Lamp Fronts are detachable to allow the cleaning of reflector and focussing of bulb.

Model No. 79—An Extra Large Heavily Silver-plated Reflector, Chromium-plated Front and Green Side Glasses are the special features of this model, which also embodies many other unique features exclusive to this New Range of Lucas Dry Battery Lamps. This model enables you to practise battery economy by means of its Dimming Switch, and still have a good beam of light, even when dimmed.

Model No. 69DSB—The popular model of the New Lucas Dry Battery Lamps has the New Dimming Switch incorporated in its design. The spring bracket prevents any jolting which is likely to harm the lamp or cause unpleasant rattle.

Battery Lamps, complete with battery and bulb, ready for use.

Code Word	No.	Finish	CASH PRICE EACH
LEECT	79	Ebony Black, with Dimming Switch and Green Side Glasses. Diam. of Front, 3¼ in.	5/-
LEEDA	69DSB	Ebony Black, with Dimming Switch. Diam. of Front, 2½ in.	4/3
FASUB	252	Spare M.E.S. Bulb (2.5 volt, .2 amp.)	3½d.
LADIM	69R	Dependable Refill ("Leadership" series)	8d.

NO. 79

NO. 79 BACK VIEW

NO. 69DSB

JOSEPH LUCAS LTD., BIRMINGHAM, 19, ENGLAND

12

Joseph Lucas Ltd 1934 (9)

---

long after Price is forgotten.

KING OF THE ROAD

Continued

Lucas Ebony Black Finish both inside and out, non-corrosive and weatherproof. All bright parts are Chromium-plated.

Secure Case-locking Clip which will not vibrate loose. Models 69D and 69 have a sensible carrier-handle.

Lucas "Leadership" Dry Battery which is supplied with each lamp provides a dependable current supply. The great recuperative power of these batteries guarantees a very considerable number of total hours' light before a refill is necessary.

Neat and compact in design. Lucas Battery Lamps are free of unnecessary projections and are therefore easy to clean.

No Loose Parts to rattle, shake off or go wrong.

Excellent Finish and Workmanship—a Lucas "King of the Road" production throughout.

Model No. 69D—Here is a lamp with all the unique features of the reputable No. 69 Model, and, in addition, has the same Dimming Switch as fitted to the more expensive models. As will be seen from the illustration, it has a secure case-locking clip and sensible handle which enables the lamp to be used as a general-purpose lamp if required.

Model No. 69—The switch is of the throw type, and so is definitely "on" or safely "off" with no risk of exhausting the Battery.

Fitted with a strong spring-grip socket for standard cycle bracket and a sensible carrier-handle this model has been in great demand during the season.

Battery Lamps, complete with battery and bulb, ready for use.

Code Word	No.	Finish	CASH PRICE EACH
LEEFE	69D	Ebony Black, with Dimming Switch (Diam. of Front 2½ in.)	3/9
LADEB	69	Ebony Black (Diam. of Front 2½ in.)	3/2
FASUB	252	Spare M.E.S. Bulb (2.5 volt, .2 amp.)	3½d.
LADIM	69R	Dependable Refill ("Leadership" series)	8d.

LUCAS
BATTERY
CYCLE .
LAMPS.

Nᵒˢ 69D & 69

NO. 69D
FRONT VIEW

NO. 69D
BACK VIEW
SHOWING SWITCH

NO. 69
BACK VIEW

NO. 69

JOSEPH LUCAS LTD., BIRMINGHAM, 19, ENGLAND

13

Joseph Lucas Ltd 1934 (10)

LUCAS "King of the Road" DYNAMO SETS

506 SET 25/-

503 SET 23/6

C25PU
TWO YEARS' GUARANTEE

DYNAMO
This is the master Dynamo supplied in all four Sets

504 SET 21/-

505 SET 23/6

The Finest Light-Givers

The Lucas range is an advance in thoughtful practical design, quality manufacture and lighting efficiency. These splendid sets embody headlamps of outstanding attractiveness, introduced only after considerable testing and development.

HEADLAMPS. Long, streamline bodies make these lamps very distinctive. Finished in All Chromium or Ebony Black, with a deep, graceful Chromium Plated Rim.

Special features include the **Patent Rubber Shock-Absorbing Bracket** actually incorporated in the body of the lamp. This absorbs vibration, eliminates rattle and provides a firm yet flexible mounting.

Lamp bodies are made of **high-quality brass**, and this, combined with the special rubber bracket, makes them absolutely **rustless**.

The alternative fixing bracket includes an ingenious non-slip locking arrangement, so that, once set, the lamp cannot possibly tilt down due to vibration.

Features common to these lamps are : 3-position Dimming Switch ; main and pilot bulbs ; large silver-plated reflector ; green side glasses in rim ; domed "Difusa" front glass ; space for dry stand-by battery.

A UNIQUE FEATURE

Cut-away view showing the Rubber Shock-Absorbing Bracket built into the body of the 506 and 503 lamps, reducing vibration to a minimum. Good contact is made certain by a silicon bronze earthing strip which runs from top to bottom of the lamp. High-quality rustless brass is the body metal of these new headlamps.

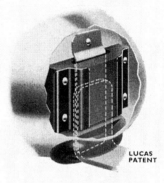

LUCAS PATENT

For Dry Batteries and Spare Bulbs, see page 15.
For Prices of Components, see separate Trade Price List.

DYNAMO. Type C25PU has Two Years' Guarantee. It is an extremely powerful 6-volt type, compact and totally enclosed, strongly made for reliability, robust from end to end. Special precautions are taken to exclude dust and mud from the bearings. Provides a brilliant steady light at all speeds. Fitted with **Lucas "Nifal" Magnets**, giving equal strength with less weight.

Output has been proved by laboratory tests to be constant over a wider range of speeds than other makes, ensuring maximum power without overloading of bulbs, and adequate light even at walking pace. A sturdy **Universal Bracket** is fitted, and the **Trigger Action release** mechanism enables the rider to engage the dynamo with the tyre quickly and easily.

An efficient **Rear Lamp** is incorporated very neatly in the base of the dynamo, having our well-known "**Realite**" **reflecting glass** (N.P.L. Certificate No. 154) which also acts as an approved reflector.

No. 506 SET. Comprises No. 506 All Chromium Plated Headlamp (rubber bracket incorporated), Dynamo No. C25PU 6v. (2 Years' Guarantee) with Trigger Type Bracket, Rear Lamp (with "Realite" Reflecting Glass) in Dynamo, Cable and Bulbs. Supplied less Battery. *Code Word—Labyc.* **25/-**

No. 504 SET. As No. 506 Set but with No. 504 Headlamp, Ebony Black, Chromium Plated Rim (with non-slip locking Metal Bracket). *Code Word—Lopia.* **21/-**

No. 503 SET. As No. 506 Set but with No. 503 Headlamp, Ebony Black, Chromium Plated Rim (rubber bracket incorporated). *Code Word—Labec.* **23/6**

No. 505 SET. As No. 504 Set but with No. 505 All Chromium Plated Headlamp (Metal Bracket). *Code Word—Lopak.* **23/6**

ever made for Cycles

Joseph Lucas Ltd 1940 (1)

Joseph Lucas Ltd 1940 (2)

Road" DYNAMO SETS

NEW No. 486 SET. A 6-volt De-Luxe Lightweight Set designed for cyclists who wish to have the best lightweight dynamo set obtainable.

The new No. 486 Headlamp is a de-luxe all-chromium plated model with a fine, graceful streamline body. The large silver-plated reflector gives an excellent beam, "Difusa" front glass ensuring width and evenness of light. Green side glasses.

No. CD18 Lightweight 6-volt Dynamo is light yet robust and offers very little drag resistance. The new Trigger Action Release Mechanism enables the rider to engage dynamo with tyre quickly and easily. New Universal Bracket.

Supplied complete with cables and bulb.

Code Word—Lopuz.

486 SET
15/-

CD18 SET
12/6

No. CD18 SET. This 6-volt set is a lightweight model for cyclists who want something lower in price than the standard "King of the Road" models, but of Lucas quality and reliability and therefore superior in lighting power and dependability to cheap sets previously available.

Dynamo CD18 is small, light, and offers very little drag resistance. The new Trigger Action Release Mechanism enables the rider to engage dynamo with the tyre quickly and easily. New Universal Bracket. **No. 586** is the Headlamp—this has a large silver-plated reflector which gives a good beam, domed "Difusa" front glass, green side glasses, etc.; finished in ebony black with chromium-plated rim. Supplied complete with cables and bulb.

Code Word—Lopmy.

For Spare Bulbs, see page 15.
For prices of Components, see separate Trade Price List.

REAR LAMP No. CT31 can be supplied for the above sets, if specially requested. It is a neat lamp, with facet type glass giving a good red light, and has a universal fixing.
Code Word—Lopal 1/3

8

LUCAS "KING OF THE ROAD"

No. 305RS

This range, with many exclusive features, and of excellent finish and workmanship, sets a new and unrivalled standard of excellence. Every lamp body is substantially made of heavy-gauge British pressed steel, very strongly riveted and welded together.

A particular feature of most models is the exclusive **PATENTED SHOCK-ABSORBING RUBBER BRACKET AND RUBBER BATTERY CUSHION IN ONE PIECE**, an outstanding development in Dry Battery Cycle Lamps. The unique features of the design are that it absorbs road vibration and eliminates rattle, ensures good contact, lengthens battery and bulb life, and provides a firm yet flexible mounting. *(Continued opposite)*

No. 305RS is a fine powerful model with two batteries. Has the patented shock-absorbing rubber bracket and rubber battery cushion in one piece, which insulates battery and bulb from road vibration. A large, easily operated switch controls the brilliant riding light and the dimmed light. Handsome and strongly made, with side glasses; supplied complete with batteries. Spare bulb clipped inside lamp body. "Difusa" glass. 8 volts, 0.2 amp.

With Shock-Absorbing Rubber Back.

Code Word
Lopow ... Ebony Black, Chromium Plated Rim ... 6/11

"Leadership" Battery Refills No. 305R, **5d.** each.

No. 310RS

No. 310RS is similar to No. 305RS model, but the base of the lamp is enlarged to house one large battery of higher capacity. It also has a fluted front glass instead of "Difusa" glass.

With Shock-Absorbing Rubber Back.

Lopod ... Ebony Black, Chromium Plated Rim ... 6/9

"Leadership" Battery Refill No. 69R, **8d.** each.
Spare Bulb No. C253 (2.5 volts, 0.3 amp.), **3d.**

Rubber Battery Cushion
Lengthens battery and bulb life and
ensures good contact

THE PATENT RUBBER BRACKET has a projection into the interior of the lamps which acts as a cushion for the battery. The illustration shows that this is part of the strong construction, and results in firmness with resilience. The battery is firmly held and prevented from rattling, good contact is ensured, while battery and bulb life is naturally lengthened.

The Finest Range—

Battery Lamps

9

Lucas Dimming Switches are positive, easy to operate, and save current.

Lucas Reflectors are accurately designed, heavily Silver Plated and non-rusting.

Other special features are **Lucas "Leadership" Dry Batteries,** ensuring dependability and long life ; bold, detachable **Chromium Plated Lamp Fronts** ; the famous **Lucas Ebony Black Finish,** weatherproof and non-corroding; very **Strong Clips** for cases and rims; particularly handsome and neat outline, and therefore **easy to clean.**

No. 311

No. 311. Handsome, powerful lamp with deep chromium plated rim and long cleanly shaped body. Dimming device incorporating easily handled 3-way switch giving "Dim," "Off" and "Bright." Green side glasses. 69R Battery.

Code Word
Liwoh ... Ebony Black, Chromium Plated Rim ... 5/11

"*Leadership*" *Battery Refill No. 69R,* **8d.** *each.*

Spare Bulb No. C253 (2.5 volts, 0.3 amp.), **3d.**

No. 79RS

No. 79RS is a large handsome model with extra big Silver Plated Reflector and Chromium Plated Front—both non-rusting. Fitted with diamond-shaped dimming switch which is very easily operated. With side glasses, battery and bulb, ready for use. Fitted with the patented shock-absorbing rubber bracket and rubber battery cushion in one piece.

With Shock-Absorbing Rubber Back.

Lopul ... Ebony Black, Chromium Plated Rim ... 5/6

No. 67, NEW MODEL, is a most useful lamp with a large, bold front, a fixed metal back socket instead of the rubber back, a neat "On-Off" Switch instead of the dimming switch, and is also fitted with a fluted glass. Another feature is the useful carrying handle which enables this model to be used as a hand-lamp, if required. This handle clips back out of the way when not in use.

With Metal Socket Back.

Lopoy ... Ebony Black, Chromium Plated Rim ... 3/6

"*Leadership*" *Battery Refill No. 69R,* **8d.** *each.*

Spare Bulb No. C253 (2.5 volts, 0.3 amp.), **3d.**

No. 67

—the Widest Choice

Joseph Lucas Ltd 1940 (5)

LUCAS "KING OF THE ROAD"

Battery Lamps

10

No. 69DRS

No. 69DRS has a one-piece front with a chromium plated rim, and diamond-shaped dimming switch which is very easily operated. Has the Shock-proof Rubber Bracket and Rubber Battery Cushion. Complete with battery and bulb.

With Shock-Absorbing Rubber Back.

Code Word
Lopye ... Ebony Black, Chromium Plated Rim ... 4/3

No. 69RS. Similar to No. 69DRS, but with "On-Off" Switch instead of the dimming switch.

With Shock-Absorbing Rubber Back.

Lopab ... Ebony Black, Chromium Plated Rim ... 3/11

No. 69

No. 69. Similar to 69RS, but with unsprung metal bracket. A useful carrier handle enables it to be used as a handlamp. With battery and bulb.

With Metal Back Socket.

Ladeb ... Ebony Black, Chromium Plated Rim ... 3/3

For above lamps | "*Leadership*" *Battery Refill No. 69R,* **8d.** *each.*
| *Spare Bulb No. C253 (2.5 volts, 0.3 amp.),* **3d.**

No. 68

No. 68. A powerful reliable lamp at a very low price. Embodies all the well-known Lucas features and has a polished silver plated reflector, positive control switch (screw-down pattern), and plated carrying handle which can be locked in position on the lamp bracket when not in use. Lamp body has attractively rounded sides.   2.5 volts, 0.3 amp.

Liwuz ... Ebony Black, Chromium Plated Rim ... 2/6

"*Leadership*" *Battery Refill No. 69R,* **8d.** *each.*

No. 48 COMBINED ELECTRIC TAIL LAMP AND APPROVED REFLECTOR. Gives a "live" red light to the rear, while the glass is the special "Realite" Reflector (awarded N.P.L. Certificate 154). Long-life battery. Even if the rider forgets to switch on, he is still protected by the reflector. Strong and well-made. Switch action is part of the lamp body. A large rubber base-pad prevents battery rattling and ensures good contact.   (2.5 volts, 0.3 amp.)

Lopax ... Ebony Black, Chromium Plated Rim ... 2/3

"*Leadership*" *Battery Refill No. TUI,* 3½d. *each.*

No. 48
Combined Tail Lamp and
Rear Reflector

The Memory of Quality remains

Joseph Lucas Ltd 1940 (6)

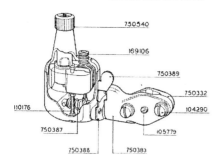

Service Ref. No. 75052.

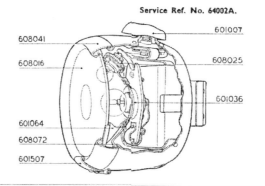

Service Ref. No. 64002A.

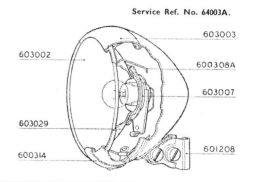

Service Ref. No. 64003A.

DESCRIPTION	PART No.	PRICE	
Pulley Assembly (including Lock Nut, Collett and Washers)	750540	1/-	each
*Bracket and Clip Assembly	750383	2/6	,,
*Trigger Assembly (including Trigger, Spring and Pivot Post)	750389	9d.	,,
*Catch Plate Assembly (including Fixing Screws and Washer)	750388	9d.	,,
*Drive Tension Spring Assembly (including Anchor Pin)	750387	6d.	,,
*Trigger Spring	750344	2/-	dozen.
Body Through Bolt	110176	} 4d. set.	
Nut for Through Bolt	169096		
Terminal Nut	169106	1/4	dozen.
*Bracket Screws	194290	2/-	,,
*Bracket Half Clip	750332	8d.	each.
*Earthing Screws	105779	1/-	dozen.
*Anchor Pin	750347	1/-	,,
Tension Spring Sleeve	750646	2/-	,,

*These items are also used on the No. C.25 P. Dynamo

DESCRIPTION	PART No.	PRICE	
Glass (Diffused)	608016	8d.	each.
Glass Retaining Wires	608072	1/-	dozen.
Front Rim	608041	2/6	each
Reflector	601064	2/-	,,
Main Bulbholder Assembly	601036	1/6	,,
Pilot Bulbholder	608025	3/6	dozen.
Front Clip and Pin	601507	6d.	each.
Switch Knob Assembly	601007	1/-	each
Switch Arm	6A-308	1/-	dozen.

The following parts are also available for Dynamo Headlamps Nos. 503/4/5/6

DESCRIPTION	PART No.	PRICE	
Glass	603407	1/-	each.
Main Bulbholder Assembly	603418	1/9	,,
Pilot Bulbholder	608025	3/6	dozen.
Bracket Assembly (504/5 Lamps only)	601208	1/9	each.

DESCRIPTION	PART No.	PRICE	
Glass	603002	4d.	each.
Glass Retaining Wire	600308	1/6	dozen.
Front Rim	603003	2/-	each.
Reflector	603029	1/9	each.
Bulbholder Assembly	603007	1/-	,,
Bracket Assembly	601208	1/9	,,
Front Clip and Fixing Wire	600315	5/-	dozen.

SPARE BULBS		DRY BATTERIES	
These are Genuine Lucas Tested type.		The well-known and reliable Lucas "Leadership" series	
No. C253 (2.5 v. 0.3 a.) for Battery Lamps	Plus P.T. 3½d. 1d.	No. 69R Large Twin-Cell for Front and Tail Lamps 9d.	
No. C252 (2.5 v. 0.2 a.) for Battery Lamps	3½d. 1d.		
No. C3515 (3.5 v. 0.15 a.) for Dynamo Tail Lamps	3½d. 1d.	No. 305R Flat Standby type for Dynamo Headlamps... 7d.	
No. C 202 for CD18 Dynamo Headlamps	1/- 3d.		
No. C43D (4 v. 0.3 a.) for Headlamp Pilot Bulb	3½d. 1d.	No. TUI Unit-Cell U2 type, for Tail Lamps ... 4d.	

Prices in this list, operative from 1st Sept., 1946, are subject to revision without notice.

Joseph Lucas Ltd 1946

PROJECTEURS  pour CYCLES

Nº 1787
Projecteur 50 m/m, verre de 50 m/m
Suspension à socle
pour garde-boue. Chromé

Nº 1730
Projecteur 85 m/m, verre de 80 m/m
Nickel 2 faces.
Suspension à griffes pour lanterne

Nº 1916
Projecteur 63 m/m, chromé
verre de 61 m/m
Suspension à griffes

Nº 1918
Projecteur 63 m/m, chromé
Verre de 61 m/m. Suspension
à rotule pour porte-lanterne

Nº 1919
Projecteur 63 m/m, chromé
Verre de 61 m/m. Suspension
à rotule pour garde-boue

MODÈLES AÉRODYNAMIQUE TRÈS PROFILÉS, 63 m/m, VERE DE 61 m/m, CHROMÉ AVEC FLAMME

Nº 1956
Suspension à griffes pour
porte-lanterne

Nº 1957
Suspension à socle pour
garde-boue

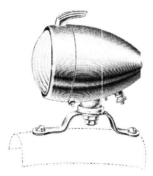

Nº 1959
Suspension à rotule pour
garde-boue

Nº 1958
Suspension à rotule pour
porte-lanterne

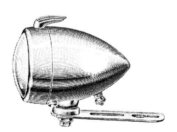

Nº 1960
Suspension à rotule avec
porte-lanterne pour expandeur

— 1 —

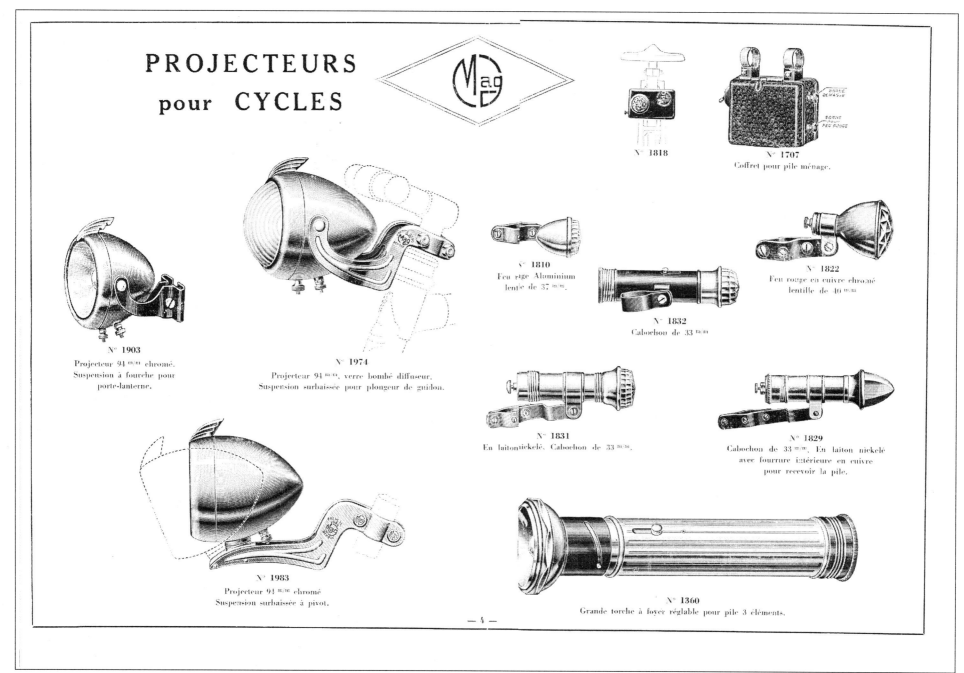

PROJECTEURS pour CYCLES

N° 1818

N° 1707
Coffret pour pile ménage.

N° 1903
Projecteur 94 m/m chromé.
Suspension à fourche pour
porte-lanterne.

N° 1974
Projecteur 94 m/m, verre bombé diffuseur.
Suspension surbaissée pour plongeur de guidon.

N° 1810
Feu rage Aluminium
lentille de 37 m/m.

N° 1832
Cabochon de 33 m/m

N° 1822
Feu rouge en cuivre chromé
lentille de 40 m/m

N° 1831
En laiton nickelé. Cabochon de 33 m/m.

N° 1829
Cabochon de 33 m/m. En laiton nickelé
avec fourrure intérieure en cuivre
pour recevoir la pile.

N° 1983
Projecteur 94 m/m chromé
Suspension surbaissée à pivot.

N° 1360
Grande torche à foyer réglable pour pile 3 éléments.

— 4 —

═══ MANDAW ═══

Electric Cycle Lamps

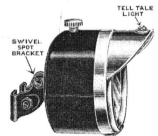

GUIDEU No 787.B

BRITISH ELECTRIC CYCLE LAMP

Registered Design No. 791,556.

Black finish with Chromium Plated Mounts. The Hood is fitted with tell-tale light, and the swivel Lamp Bracket can be adjusted to project light at any angle. Size of door 4 in. weight complete with Twin-Cell Battery, 21 ozs. Sold complete with Special Guideu Twin Battery and Special Spotlite Bulb.

5/6 complete

REPLACEMENT BULBS.
No. **366M** Bulb for Guideu Cycle Lamps.
4½d. each

REPLACEMENT BATTERIES.
No. **333B** Mandaw Battery for Guideu Lamps.
7d. each.

═══ MANDAW ═══

Page 1

═══ MANDAW ═══

Electric Tail Lamps

MANDAW. REAR LAMP. No 123.B.
PATENT. No 429400.

Mandaw No. 123B.

BRITISH MADE REAR LAMP

(Patent No. 429,400.   Licence No. 349, 347)

Black finish with Plated Mounts. Clip especially designed to fit oval, "D" and round stays. Unbreakable red projector, especially designed to project red light to the rear and sides. Fitted with low consumption bulb and supplied complete with Unit Cell

1/- complete

"MANDAW" REPLACEMENT BATTERIES.
No. **533B.**
Size 2⅜ × 1⅜ in. Single Cell Specially constructed for Tail Lamps.

533B

3d, each.

═══ MANDAW ═══

Page 2

Mandaw Co Ltd 1935 (1)

Mandaw Co Ltd 1935 (2)

Electric Tail Lamps

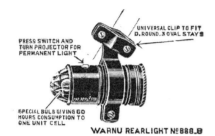

PRESS SWITCH AND
TURN PROJECTOR FOR
PERMANENT LIGHT

UNIVERSAL CLIP TO FIT
D. ROUND. & OVAL STAYS

SPECIAL BULB GIVING 60
HOURS CONSUMPTION TO
ONE UNIT CELL

WARNU REARLIGHT No 888.B

"Warnu" No. 888B.

BRITISH MADE REAR LAMP.

Registered No. 791,557.
Patent No. 429,400.

This perfect rear lamp can be supplied in Black or Ivory White Finish. The clip is specially designed to fit oval, D, and Round Stays. Red facetted projector specially designed to project Red light to rear and sides. Fitted with special "M" Low Consumption Bulb, giving 60 hours consumption to one unit cell.

Fixed Retail Price

2/6 complete

LOW CONSUMPTION REAR LAMP BULBS, GENUINE 'M' MAKE.

No. 666M.

Gives 60 hours consumption to one Single Unit Cell.

4½d. each.

No. 666M

Mandaw Co Ltd 1935 (3)

MI FOCUSSING 6 VOLT LAMP SET

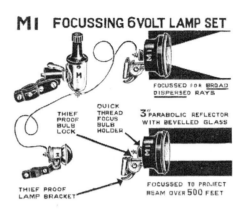

FOCUSSED FOR BROAD
DISPERSED RAYS

THIEF
PROOF
BULB
LOCK

QUICK
THREAD
FOCUS
BULB
HOLDER

3" PARABOLIC REFLECTOR
WITH BEVELLED GLASS

THIEF PROOF
LAMP BRACKET

FOCUSSED TO PROJECT
BEAM OVER 500 FEET

M.I. DYNAMO SET.

Front and Rear Set complete with Genuine "M" Dynamo Bulbs. (First quality Foreign Manufacture).

Specification:

6 volt Chromium Plated Dynamo.

Black Focus Searchlight with 3 in. Front and Rustless Thief-proof Lamp Bracket.

Chromium Plated Rear Lamp with 1½ in. Facetted Glass.

Two full-length best quality Insulated Cables, with Terminal Clips attached.

15/9 per set complete

Mandaw Co Ltd 1935 (4)

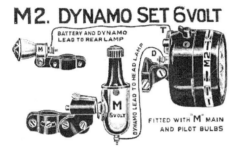

M2. DYNAMO SET 6 VOLT

M.2. DYNAMO SET.

Front and Rear Set complete with Genuine "M" Dynamo Bulbs. (First quality Foreign Manufacture).

Specification:

6 volt Chromium Plated Dynamo.

Black and Chrome Head Lamp with $3\frac{1}{2}$ in. Door, fitted with main and "Pilot" Bulbs. Supplied with Terminal for Battery and Dynamo Lead to Tail Lamp

Black Rear Lamp with 1 in. Facetted Glass.

Two Full length best quality Insulated Cables with Terminal Clips attached.

16/9 per set complete

Mandaw Co Ltd 1935 (5)

M3. DYNAMO SET 4 VOLT

THIS SET IS FITTED WITH GENUINE "M" DYNAMO BULBS

M.3. DYNAMO SET.

Front and Rear Set complete with Genuine "M" Dynamo Bulbs. (First quality Foreign Manufacture).

Specification:

4 volt Chromium Plated Dynamo.

Black Headlamp with $3\frac{1}{2}$ in. Door and Rustless Spring Bracket, fitted for use with auxiliary PL Battery.

Black Rear Lamp with 1 in. Facetted Glass.

Two full-length best quality Insulated Cables with Terminal Clips attached.

12/- per set complete

Mandaw Co Ltd 1935 (6)

━━ MANDAW ━━

M4 DYNAMO SET 6 VOLT

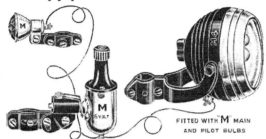

FITTED WITH "M" MAIN
AND PILOT BULBS

M.4. DYNAMO SET.

Front and Rear Set complete with Genuine ''M'' Dynamo Bulbs. (First Quality Foreign manufacture).

Specification:

6 volt Chromium Plated Dynamo.

Black and Chromium Projector with $4\frac{1}{2}$ in. Door and Focussing Bracket, for use with auxiliary PL Battery. The 4-way switch is operated as follows:

'D.D.' for Dim Dynamo.
'D.B.' for Dim Battery.
'B.B.' for Bright Battery.
'B.D.' for Bright Dynamo.

Chromium Rear Lamp with $1\frac{1}{2}$ in. Facetted Glass.

Two Full Length Best quality Insulated Cables with Terminal Clips attached.

17/9 per set complete

━━ MANDAW ━━

Page 7

Mandaw Co Ltd 1935 (7)

━━ MANDAW ━━

Genuine 'M' Dynamos for Cycle Lamps.

No. 110. 6 volt Genuine 'M' DYNAMO.

Chromium Plated Body with Black Mounts for Fitment on Front or Rear Wheels.

10/- each

No. 111. 4 volt Genuine 'M' DYNAMO.

Chromium Plated Body with Black Mounts for fitment on Front or Rear Wheels.

10/- each

━━ MANDAW ━━

Page 9

Mandaw Co Ltd 1935 (8)

━━━━━━━━━━ MANDAW ━━━━━━━━━━

Genuine "M" Dynamo

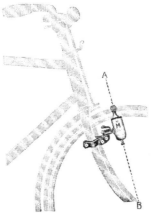

This Dynamo has a Chromium Plated Casing, with Black Mounts, and is supplied with a clip suitable for fitting to all standard size front forks and rear stays, and can be operated by means of a foot or hand lever.

The Rotor is carried on a double row of ball races, which are specially packed with a very super quality non-diminishing lubricant. It is therefore not necessary to supply further lubrication, the machine when sent out is ready for long use without any attention.

Instructions for Fitting.

Dynamos can be fitted on either front or rear wheels but they must be attached to the right hand side of the Cycle, reversed running is not advisable.

The Dynamo must be fixed in such a manner that the top wheel and bottom terminal screw are exactly in line with the centre of the hub of the wheel. The top driving wheel should touch the Tyre at the centre of the wall, so as to avoid wear of the cover.

The contact screw in the clip must be screwed right home through the enamel.

━━━━━━━━━━ MANDAW ━━━━━━━━━━

Page 8

Mandaw Co Ltd 1935 (9)

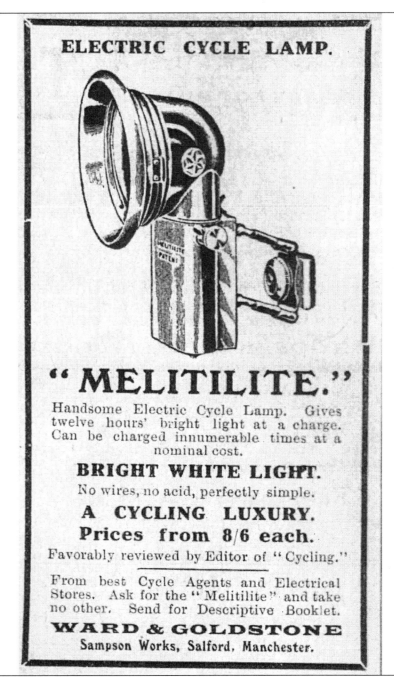

Melitilite advertisment of 1908

MILLER CYCLE DYNAMO SET.

The Miller Cycle Dynamo Set embodies all the good points necessary to make this form of lighting successful. Full details are given in separate booklet, which will be forwarded on application.

The lamp is supplied complete with bracket, clips, and wiring, ready for attachment to bicycle, an operation which can be performed in a very few minutes.

Price :
No. 56C.D.,
Head Lamp and Dynamo,
£1 2 0

Cable Code: 00110.

No. 57C.D.,
complete with Tail Lamp,
£1 5 0

Cable Code: 00111.

The Lamps that won't go out

23

H. Miller & Co Ltd 1923 Their first dynamo system

Miller Electric Cycle Lamp Set

SPECIFICATION
*HEAD LAMP—3¼in.
DYNAMO.
CABLE FOR WIRING.
FINISH—Ebony, parts plated*

56.C.A
Cable Code: 00305
£1 4 6

*HEAD LAMP—3¼in.
DYNAMO.
TAIL LAMP.
CABLE — For wiring both lamps.
TWO CLIPS—For attaching Cable to frame.
FINISH—Ebony, parts plated.*

57.C.A
Cable Code: 00306
£1 7 6

Exhaustive experiments by the MILLER electrical experts have resulted in what is undoubtedly the finest electric cycle lamp ever manufactured. It is light, compact, and gives a beam entirely superior to any other lamp on the market.

There is no flicker, no slipping of the pulley, and the lamp cannot be overcharged no matter what the speed. Such a set requires no attention whatever. It is always ready, unaffected by weather conditions, is fixed in a moment to the machine, and will wear for many years.

It has a detachable glass, SILVER - PLATED parabola reflector, adjustable bulb holder, and is supplied ready wired for fitting, together with full instructions and sectional drawings.

58 C.A. 6/-	59.C.A. 3/-
Cable Code : 00307	Cable Code : 00308
HEAD LAMP only.	TAIL LAMP only.

60.C.A. 18/6
Cable Code : 00270
DYNAMO only.

The spare parts to this model will be found on page 27.

4

H. Miller & Co Ltd 1927

Cycle Dynamo Electric Lighting Set

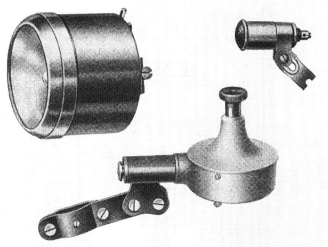

We have now re-designed our well-known Cycle Dynamo Set, and the new pattern has already received a flattering reception. It combines all the good qualities of our previous sets, and has, in addition, several new features which will render it even more attractive than our old patterns.

The dynamo is of the rotary magnet type and generates current with the minimum of friction. Our old principle of driving from the tyre is retained, and this dynamo will be found to maintain all the non-slipping qualities of its predecessors.

The Head lamp is also very attractive, being fitted with a 3in. glass and heavily silvered plated reflector. The switch is positive three-way Off-Dynamo-Battery. The bulb is of the highest focal quality for dynamo work, and the bulb holder is adjustable and of the standard screw-in pattern.

The lamp has a receptacle in which a flash lamp battery is carried so that when the machine is standing and it is desired to have a light it can be switched over on to the battery, thus avoiding what in the past, has been one of the chief disadvantages of self-generating sets.

The whole finish of the production is perfect. The head and tail lamp in our well-known cellulose ebony with nickel plated parts, while the Dynamo is satin crystaline finish.

SPECIFICATION

HEAD LAMP	.	
DIAMETER	.	4in.
GLASS	.	3in
SWITCH	.	3-way
BULBS	Head 4v. 3a.,	Tail5·5 v. ·15a.
REFLECTOR	.	Silver-plated
WEIGHT, less battery	.	2lbs. 1oz.
BRACKET	.	side screw
BATTERY	Standard flash lamp	

56 D.B. Price With Bulb **19/-**
Cable Code - 00360.

57D.B. Same set with tail lamp **21/-**
Cable Code - 00381.
Battery 6d. each extra.

52D.B. Head Lamp only **6/-**

60D.B. Dynamo only - **14/-**

59D.B. Tail Lamp only **2/-**

For Spare Parts for above, see page 61.

40

Cycle Electric Battery Lighting Sets

No. 555   No. 555

THE MILLER BATTERY LAMP No. 555

This battery lamp de-luxe is of unique and attractive appearance, as will be seen from the illustration. It is fitted with a neat handle, which is secured when not in use by a safety catch which renders it proof against any vibration. The back is strong and is rivetted to the body of the lamp thus rendering any breakage in this direction impossible.

The glass and reflector (which is heavily silver-plated) are of large size thus giving a first-class light. The focal bulb is of the highest quality, and the whole lamp is finished in our well-known cellulose ebony finish with plated switch, handle and front.

The lamp as a whole makes an attractive addition to any high-class bicycle, and it is equally useful as an inspection lamp or for household use. It is designed to accommodate a high quality two cell battery. The switch is three way On and Off and has an additional terminal which permits a dynamo to be used if desired.

SPECIFICATION

HEIGHT, minus handle	4in.
GLASS	3in.
SWITCH	2 way
BULB	2.5 volt. 3 amp.
REFLECTOR	heavily silver-plated
FOLDING HANDLE	
WEIGHT, without battery	
BRACKET	side screw
Designed to accommodate two-cell battery.	

555 Price - 7/-
With Bulb and Battery.
Cable Code : 00387.

555 Price - 6/-
Less Battery.
Cable Code : 00379.

BATTERY B23 - 1/- each.

41

H. Miller & Co Ltd 1930 (1)   *H. Miller & Co Ltd 1930 (2)*

Battery Lighting Set for Cycles

No. 333

The No. 333 is a low priced but very strong and reliable battery lamp of ample capacity, and is fitted with a focal bulb of highest quality. The switch is definite and strong. Will not get out of order. The lamp is stormproof. Driving rain cannot find its way into the interior, while the whole lamp is finished in best possible style—finest ebony black, with nickel-plated parts and silver-plated reflector. The folding handle is fastened when not in use by an ingenious and very strong clip.

SPECIFICATION	No. 333 Price 3/3
	With Bulb and Battery.
	Cable Code: 00388.
HEIGHT (minus handle) . 4½in.	
WEIGHT (less battery) . 9oz.	No. 333 Price 2/3
FRONT GLASS . . 2½in.	With Bulb, less Battery.
BULB FOCAL . . 2·5 volt.	Cable Code: 00382.
FOLDING HANDLE.	
	BATTERY No. B23 ... each 1/-

42

Battery for Cycle Lamps

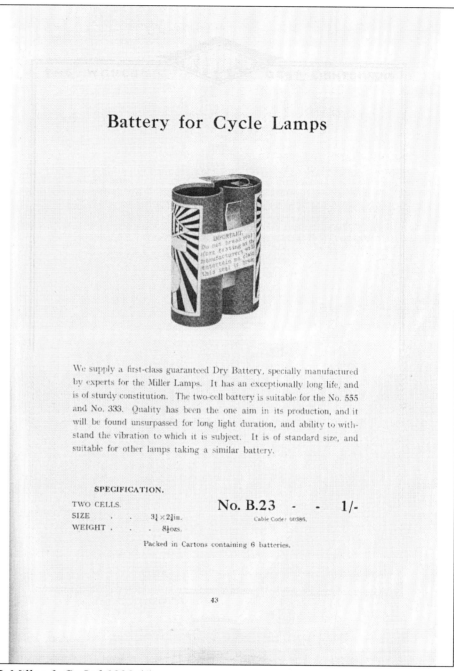

We supply a first-class guaranteed Dry Battery, specially manufactured by experts for the Miller Lamps. It has an exceptionally long life, and is of sturdy constitution. The two-cell battery is suitable for the No. 555 and No. 333. Quality has been the one aim in its production, and it will be found unsurpassed for long light duration, and ability to withstand the vibration to which it is subject. It is of standard size, and suitable for other lamps taking a similar battery.

SPECIFICATION.	No. B.23 - - 1/-
TWO CELLS.	
SIZE . . 3¼×2½in.	Cable Code: 00386.
WEIGHT . . 8½ozs.	

Packed in Cartons containing 6 batteries.

43

H. Miller & Co Ltd 1930 (3)   *H. Miller & Co Ltd 1930 (4)*

Miller Cycle Dynamo Set

(6 Volt)

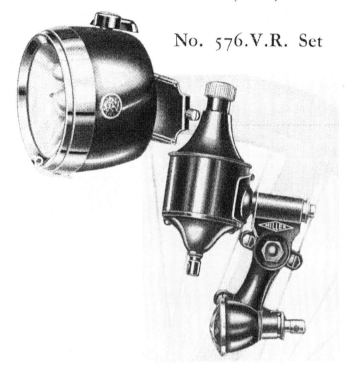

No. 576.V.R. Set

No. 576.V.R. Set

THE No. 576.V. Set is for front fork fitting, while the No. 576.V.R. is similar to the No. 576.V., but is constructed to operate on the back wheel. By means of the Miller patented bracket, the dynamo can be mounted in a secure and extremely neat manner. The Tail Lamp is also mounted on the dynamo bracket, the whole set being of the utmost reliability, very great light-giving power, neat yet accessible, in fact, the latest word in Cycle Dynamo Sets. Made in three types of fitting, viz., for D or round stays, or for machines fitted with calliper brakes. If specially ordered, it can also be had with cables suitable for tandems, costing sixpence extra.

SPECIFICATION

HEAD LAMP.	Reflector	Broad beam, 3¼in. diameter.
No. 586.V.	Glass	3¼in.
	Front	Chromium-plated, 4in. diameter.
	Bulbs	Main (Dynamo) 6v. .5a.
	"	Pilot (Battery) 4v. .3a.
	Weight (less Battery)	19 oz.
	Finish	Ebony, with Chromium parts.
DYNAMO.	Capacity	6 volt.
No. 606.V.R.	Bracket	Rear Wheel.
	Height	5½in.
	Weight	1lb. 10 oz.
	Finish	Ebony with Chromium Parts.
TAIL LAMP.	Bulb	8v. .1a.
No. 596.V.R.	Glass	1¾in. diameter.
	Weight	2 oz.

BRACKETS. This set is supplied with brackets to fit D or round back stays or machines fitted with calliper brakes.

When ordering, please state the type of Clips required.

No. 576.V.R. Price 23/6 complete Set

Comprising Head Lamp, Tail Lamp, Dynamo, and all Cables and Clips.

No. 566.V.R. - 21/6

As No. 576.V.R., but without Tail Lamp.

Cable Code : 00699

TWENTY years' experience in the design and manufacture of Cycle Dynamo Sets is built into this remarkably efficient lamp.

Being a full 6 volt set, it gives an amazingly brilliant beam, the value of which is enhanced by the Miller "Broadbeam" Reflector, which not only maintains length of light, but also gives remarkable extra width.

Two bulbs are fitted, the main being fed from the dynamo, and the pilot from the stand-by battery which is housed in the head lamp itself.

This new set for 1935 has the following advantages :—

1. It has a bulb holder giving very fine adjustment and so permits the light to be perfectly focussed.

2. There are no loose wires in the head lamp.

H. Miller & Co Ltd 1935 (1)

Miller Cycle Dynamo Set

(6 Volt)

No. 576.V. Set

No. 586.V.

No. 606.V.

No. 596.V.

B EING a full 6 volt set it gives an amazingly brilliant beam, the value of which is enhanced by the Miller " Broadbeam " Reflector, which not only maintains length of light, but also gives remarkable extra width.

The undermentioned advantages over all others should be noted :—

1. It has a bulb holder giving very fine adjustment, and so permits the light to be perfectly focussed.

2. There are no loose wires in the head lamp.

No. 576.V. Set

T HIS 6 volt set is the product of over twenty years' experiment and experience in the construction of Cycle Dynamo Sets. It embodies all those points which have made Miller sets world famous. Being a 6 volt set, it naturally gives a very brilliant beam, having all the features and refinements of our 4 volt set, including the epoch-making Miller " BROADBEAM " REFLECTOR.

The Head Lamp has two bulbs, the main one fed by the dynamo, and the Pilot Bulb for use when the standby battery, which fits into the lamp itself, is called into operation.

SPECIFICATION

HEAD LAMP.	Reflector -	" Broadbeam," 3¼in. diameter.
No. 586.V.	Glass - - - -	3¼in.
	Front -	Chromium-plated, 4in. diameter.
	Bulbs -	Main (Dynamo) 6v. .5a.
	"	Pilot (Battery) 4v. .3a.
	Weight (less Battery) - -	19 oz.
	Finish -	Ebony, with chromium parts.
DYNAMO.	Capacity - - - -	6 volt.
No. 606.V.	Bracket - - -	Front Fork.
	Height - - -	5¼in.
	Weight - - -	1lb. 10 oz.
	Finish -	Ebony, with chromium parts.
TAIL LAMP.	Bulb - - - -	8v. .1a.
No. 596.V.	Glass - - -	1⅜in. diameter.
	Weight - - -	2 oz.

Price 23/6

Cable Code : 00674

No. 586.V.	Cycle Dynamo Head Lamp	7/6
No. 606.V.	Cycle Dynamo only -	16/-
No. 596.V.	Cycle Dynamo Tail Lamp	2/-

No. 566.V. Set as above, but without Tail Lamp 21/6

Cable Code : 00700

H. Miller & Co Ltd 1935 (2)

Miller Cycle

No. 57.V.R. Set

58.V.R.

60.V.R.

59.V.R

THE Dynamo is precisely similar in type to our No 576 models, but is in 4 volt capacity.

In every respect the set is equal in durability, equal in reliability, and of the same quality and finish.

Dynamo Sets (4 volt)

No. 57.V.R. Set

THIS 4 volt Dynamo introduced last year has proved its value, and it is recognised as a standard of excellence for this type of set.

THE HEAD LAMP is perfectly weather-proof, having the "Broadbeam" reflector and chromium-plated front. Within the lamp is space for an auxiliary battery which can be brought into operation when the dynamo is out of action (e.g., machine left standing or held up in traffic). There is also a spare bulb carrier inside the head lamp so that a replacement can be made without any delay or trouble. The switch is 3-way (Dynamo—Battery—Off), and the head lamp bracket is easily adjustable.

THE DYNAMO is very strongly made and, in addition to the voltage regulator, has a much improved pivot which comes readily into action. It is of the revolving magnet type, the armature spindle running on ball-bearings. The clip is adjustable.

THE TAIL LAMP has a 1¼in. faceted red glass, and detachable bulb holder. The clip enables the lamp to be adjusted to any angle.

SPECIFICATION

HEAD LAMP for Nos. 56.V.R. and 57.V.R.

Front Glass	-	3½in.
Bulb	-	4 volt, .3 amp.
Space for Battery in the head lamp.		
Weight (less battery)	-	13½oz.

Reflector	-	Silver Plated
Switch, 3-way, Off—Dyno—Battery.		
Bracket Clip adjustable to any position.		
Finish Ebony and Chromium-plated front.		

DYNAMO.			TAIL LAMP.		
Ball-bearing.			Bracket	-	Adjustable
Pulley	-	Non-slip.	Red Glass	-	1¼in.
Bracket	-	Front Fork	Bulb	-	4 volt, .5 amp.
Weight	-	22 oz.	Weight	-	2 oz.
Finish Ebony with Plated Nuts and Cover.					

Weight of Set No. 57.V.R. complete, 2lb. 4½oz.

No.	Cable Code	Price
56.V.R. complete - - -	00645	19/-
57.V.R. Set, complete with Tail Lamp - -	00644	21/-
58.V.R. Head Lamp only, complete with Bulb -	00666	6/-
59.V.R. Tail Lamp only, complete with Bulb -	00667	2/-
60.V.R. Dynamo only, with Bracket - -	00668	14/-

The Set is supplied complete with Cable and Cable Clips.

H. Miller & Co Ltd 1935 (3)

Miller Cycle

No. 54.V.R. Set

58.R.H.

60.V.R.

THIS Dynamo is precisely similar to that fitted to our No. 57.V.R. Set, but is adjusted to give the smaller output to supply head lamp bulb only. The head lamp will give a light equal to that of the No. 57.V.R., but is not designed to carry a stand-by battery.

Although low in price, the set is of equal durability, and the same quality of workmanship and materials is put into it as in our higher priced sets.

H. Miller & Co Ltd 1935 (4)

Dynamo Sets

(4 Volt)

No. 54.V.R. Set

THE Dynamo of this set is precisely similar to that fitted to the No. 57.V.R. Set, but the output is adjusted to serve the head lamp only. The head lamp itself is of a different type from that fitted to the No. 57.V.R. Set. The reflector is of the same size, but there is no provision made for a stand-by battery to be carried in the lamp.

The head lamp is perfectly weather-proof. It is fitted with a " Broadbeam " reflector. It is very neat, light in weight, and gives good illumination. The bulb is of the same capacity as that fitted to the No. 57.V.R. Set.

The No. 54.V.R. Set can be supplied with Tail Lamp if desired, when it is known as the No. 55.V.R. In this set the tail lamp has a 1½in. faceted red glass and detachable bulb holder. The clip enables the lamp to be adjusted to any angle.

SPECIFICATION

HEAD LAMP.	Diameter	4in.
	Reflector	Silver-plated.
	Switch	2-way—Off-On.
	Weight	7½ oz.
	Glass	3½in.
	Bulb	4v. .3a.
	Bracket Clip	Adjustable.
	Finish	Ebony.
DYNAMO.	Ball-bearing.	
	Pulley	Non-slip.
	Bracket	Front Fork.
	Weight	22 oz.
	Finish	Ebony with plated nuts and cover.

	Cable Code	Price
No. 54.V.R. Set, without Tail Lamp	00646	17/-
No. 55.V.R. Set, with Tail Lamp	00647	19/-
No. 58.R.H. Head Lamp, complete with bulb	00669	3/-
No. 59.V.R. Tail Lamp, complete with bulb	00667	2/-
No. 60.V.R. Dynamo only, with bracket	00668	14/-

Miller Battery

THIS is probably the finest cycle battery lamp ever produced, and is a de luxe model in every sense of the word. It has, of course, the new Miller Broad Beam Reflector, and its consequent advantages (see pages 68 and 69). The main and pilot bulbs save current and prevent total light failure. Though the lamp is a round pattern, it is so designed that the battery-well enables the lamp to be stood upright when not on the cycle. The new type spring back consists of an enclosed and concealed spring, rotating on a specially designed brass bush and reduces risk of breakage to a minimum. The new enamel finish gives the lamp a most attractive appearance, and the cyclist who wants the very best in battery lamps need look no further than the Miller No. 7.R.7.

SPECIFICATION

Height	-	4½in.	Reflector - - Broad Beam.	
Diameter of Front	-	4in.	Switch - - 3-way.	
Diameter of Front Glass	-	3¹⁵⁄₁₆in.	Bracket - - Side Screw.	
Bulbs—Main -	2.5 volt, .3 amp.			
Pilot -	2.5 volt, .06 amp.		Weight (less battery) - 1 lb.	

Complete with Battery and Bulbs

No. 7.R.7 Black and Chrome parts **8/6** All Chrome finish ... **11/6**

Cable Code : 00649 *Cable Code : 00671*

Replacement Battery No. B.23, **8d.** each.
Replacement Bulbs, Main No. 253, Pilot No. 251, **5d.** each.

Lamps for Cycles

THIS is a round pattern battery lamp, similar in character to our No. 7.R.7. It has the famous Miller Broad Beam Reflector (see pages 68 and 69). It is fitted with a focal bulb and an improved bulb and battery-saving dimming device (dim and bright switch). It also has our new type hasp-opening front, and the front, hasp, and switch are chrome-plated. It takes a standard Miller 2-cell battery, the specially designed well for which forms a base for the lamp to stand upright when not on the machine. The back spring rotates on a special brass bush, eliminating the danger of breakage at this point. The beautiful finish rounds off a very attractive lamp, from every point of view.

SPECIFICATION

Height	-	4⅜in.	Reflector - - Broad Beam.	
Diameter of Front	-	4in.	Switch - - 3-way.	
Diameter of Front Glass	-	3⁷⁄₈in.	Bracket - - Side Screws.	
Bulb -	- Focal, 2.5 volt, .3 amp.		Weight (less battery) - 14oz.	

No. 6.R.6 Complete with Battery and Bulb **6/9**

Cable Code : 00650

Replacement Battery No. B.23, **8d.** each.
Replacement Bulb No. 253 **5d.** each.

H. Miller & Co Ltd 1935 (5)

Miller Battery

AN entirely new MILLER line is this Double Battery Lamp, designed to take two flash lamp batteries. This enables an 8 volt lamp bulb to be used, which results in a very intense beam of light for dimly lighted streets and lanes. Current wastage, however, in brilliantly lit streets is avoided by use of the dimmer switch, conserving the current yet giving reasonably good illumination. This feature intelligently used with new batteries will prevent bulb breakage from overload.

SPECIFICATION

Diameter -	4½in.	Switch -	Three-way.
Diameter of Front -	4in.	Brackets -	Side Screw.
Diameter of Front Glass -	3¾in.	Weight (without Battery) -	15 oz.
Bulb -	8v. .1a.	Finish -	Ebony, with chromium-
Reflector, BROADBEAM, nickel-plated.			plated parts.

No. 8.R.8 Price, complete with Miller 8 volt Bulb and two Miller Batteries **7/-**

Cable Code : 00682

No. B.34. Replacement Batteries - **1od.** per pair.
No. 851. Replacement Bulbs - **5d.** each.

Lamps for Cycles

THIS is a square type battery lamp, further improved since its introduction by the addition of the Miller Broad Beam Reflector (see pages 68 and 69). It has the latest Miller hasp opening front and spring back. The provision of main and pilot bulbs ensures bulb and battery economy and practically eliminates total bulb failure, as it is most unlikely that both bulbs would burn out together, thus there is always a " get-you-home " bulb. The improved beam and the superb finish make this a first choice among cyclists whose preference is for the square type lamp.

SPECIFICATION

Height -	4in.	Reflector -	Broad Beam.
Diameter of Front -	3½in.	Switch -	3-way.
Diameter of Front Glass -	3⅛in.	Bracket -	Side Screw.
Bulbs—Main	2.5 volt, .3 amp.	Special Ventilating Chamber.	
Pilot	2.5 volt, .06 amp.	Weight (less battery) -	14oz.

No. 5.R.5 Complete with Battery and Bulbs **6/-**

Cable Code : 00651

Replacement Battery No. B.23, **8d.** each.
Replacement Bulbs, Main No. 253, Pilot No. 251, **5d.** each.

H. Miller & Co Ltd 1935 (6)

Miller Battery

AS a medium-priced cycle battery lamp, the Miller No. 4.R.4 would be hard to beat. It has several special characteristics, one of which is the Broad Beam Reflector (see pages 68 and 69). The unique Miller bulb and battery-saving dimming device is fitted, by means of which the bulb can be protected during the first few hours' use of a new battery (when so many bulbs formerly were burned out, due to the battery being of necessity over rating). Current is also saved by using the dimming switch when the cycle is standing or ridden in brilliantly lighted streets. The heavy gauge steel body is everywhere weatherproof. The detachable bulb holder has a focussing arrangement, and the lamp is supplied complete with battery and bulb.

SPECIFICATION

Height	-	4in.	Reflector	- Broad Beam.
Diameter of Front	-	3⅝in.	Switch	- 3-way.
Diameter of Front Glass	-	3¼in.	Finish - Ebony, with Chrome Front.
Bulb	- Focal, 2.5 volt, .3 amp.	Weight (less battery)	- 11½oz.

No. 4.R.4 Complete with Battery and Bulb 5/-

Cable Code : 00652

Replacement Battery No. B.23, **8d.** each.
Replacement Bulb No. 253, **5d.** each.

Lamps for Cycles

THE New No. 444 now has the Miller Broad Beam Reflector (see pages 68 and 69). Made of heavy gauge steel throughout; ebony finish; weatherproof in every possible respect; semi-hinged detachable front, 3½in., with stormproof rubber seating and screw fastening; well bottom, allowing ventilating space; highly polished silver-plated reflector, 3½in.; detachable bulb holder, with special focussing arrangement; screw-down switch, with solid brass pin (cut thread) and solid brass inserted nut (cut thread); spring back. Price includes bulb and battery. Note : — This lamp, in common with our cheaper model, No. 222, is supplied complete with the MILLER battery of full standard size and highest quality.

SPECIFICATION

Weight (less battery)	-	10½oz.	Front -	-	3½in.
Height	-	4in.	Bulb	Special Focal Filament.
Glass -	-	2⅜in.		

No. 444 Complete with Battery and Bulb 3/9

Cable Code : With Battery 00591; Without Battery 00592

Extra Batteries No. B.23, **8d.** each.
Replacement Bulb No. 253, **5d.** each.

H. Miller & Co Ltd 1935 (7)

Miller Battery

THIS is a square type lamp, inexpensive in price, and very popular among a widening circle of riders, as it has the Miller Patent Dimming Device. This novel invention enables the light to be dimmed for town use or when the machine is left standing, while the brilliant car-type beam can be used for darker streets and lanes. In addition to saving current in this way, it is also possible to protect the bulb during the first use of a new battery, which may be of necessity over rating and liable to burn out the bulb if this patent dimming device were not available. A folding handle is fitted which enables the lamp to be used for other than cycling purposes if desired.

SPECIFICATION

Height (handle folded)	4in.	Reflector	Silver Plated.
Front Glass (diameter)	2⅝in.	Switch	3-way.
Bulb	Focal, 2.5 volt, .2 amp.	Weight (less battery)	6½oz.
Handle	Folding.		

No. 3.R.3 Complete with Battery and Bulb 3/6

Cable Code : 00653

Replacement Battery No. B.23, **8d.** each.
Replacement Bulb No. 253, **5d.** each.

Lamps for Cycles

THE LAMP WITH THE WONDERFUL LENS

THIS novel production will be universally known as " the lamp with the wonderful lens." By scientific design, the result of long experience and an unparalleled knowledge of the cyclists' needs, the following features are all combined in a single lens : a long beam, a broad beam, side reflection to warn cross traffic, and upward reflection to assure the rider himself that the lamp is alight when riding in well-lighted thoroughfares. Hitherto these features, where they have been embodied, have usually necessitated different small attachments and gadgets.

The lamp is not only the most novel lamp on the market, but is the only low-priced lamp which combines all the qualities above mentioned. It takes a standard two-cell battery.

SPECIFICATION

Height	4in.	Bulb	Focal 2.5v. .3a.
Weight (less battery)	7½oz.	Folding Handle.	
Patented Front Lens	2⅜in.	Finish — Ebony, with chromium-plated front and switch.	

No. 3.L.3 Complete with Bulb and Battery 3/-

Cable Code : 00683

No. B.23.	Replacement Batteries		**8d.** each.
No. 253.	Replacement Bulbs		**5d.** each.

Miller Battery

Lamps for Cycles

THIS well-known model was the first Electric Battery Lamp to be marketed at a really popular price. Considering its cost, it has a unique specification. It is made of heavy gauge steel throughout; weatherproof in every possible respect; detachable front, with stormproof seating; well bottom, allowing ventilating space; highly polished plated brass reflector; large detachable carrying handle, eccentrically mounted to avoid fouling battery; screw-down switch, with solid brass pin (cut thread), and solid brass inserted nut (cut thread). The price includes bulb and battery.

Takes full size standard 2-cell battery.

SPECIFICATION

Height (minus handle)	4in.	Front Glass	2¼in.
Weight (less Battery)	6oz.	Bulb	Focal, 2.5 volt
	Folding Handle.		

No. 222 Complete with Battery and Bulb **2/6**

Cable Code : With Battery 00589 ; Without Battery 00590

No. 222.T.

The No. 222 Lamp is also supplied as a Rear Lamp, being precisely similar, but with red glass and special low consumption bulb.

Price Complete with Battery and Bulb **2/6**

Cable Code : 00672

Extra Batteries, No. B.23, **8d.** each.

Bulb for No. 222, **5d.** each.   Bulb for No. 251, **5d.** each.
  „   „ No. 222T, **5d.** „   „   „ No. 252, **5d.** „

THIS entirely new production for the coming season exactly fills the requirements of those who only do the smallest possible amount of night-riding and therefore require a really cheap but efficient lamp. This model takes an ordinary flash lamp battery instead of a 2-cell battery. The front and reflector are precisely similar to those fitted to our No. 222 Battery Lamp.

SPECIFICATION

Height	3½in.	Silver-plated Reflector.	
Weight	6 oz.	Finish	Ebony.
Glass	2¼in.	Bulb	4v. .3a.

No. 111 Complete with Miller Bulb and Miller Battery **2/-**

Cable Code : 00684

No. B.34.	Replacement Battery	-	**5d.** each.
No. 353.	Replacement Bulb	-	**5d.** each.

H. Miller & Co Ltd 1935 (9)

Miller Battery

THIS is an extremely interesting Head and Tail Lamp Combination Set, and its novel features will assume a widespread importance in view of the importance of an efficient rear light. The tail lamp is connected with the head lamp, and automatically becomes alight when the latter is switched on. Very little extra current is needed as compared with the head lamp alone. The set is a combination of our No. 222 Battery Lamp (see page 64) slightly adapted for this new purpose, and our Electric Tail Lamp as fitted to the cycle dynamo sets described on pages 52 and 53.

SPECIFICATION

HEAD LAMP. No. 222 (page 64) with tail lamp connection.
TAIL LAMP. No. 59.V.R. (page 53) with special Bulb.
CABLE. Best quality, silk covered.

No. 2.S.2 Complete with Bulbs, Battery, and Cable **4/3**

Cable Code : 00848

Replacement Battery No. B.23, **8d.** each.
Replacement Bulbs—Head No. 253 ; Tail No. 251, **5d.** each.

Sets for Cycles

THE Miller No. 666 was introduced to meet the demand for a lamp with a separate container — which may be attached to the top tube of the cycle and connected to the head lamp by cable. The round head lamp is particularly neat in appearance and is as fitted to our Dynamo Set (No. 54.V.R.). The battery container takes a standard 2-cell battery, is entirely weatherproof, and light in weight.

SPECIFICATION

HEAD LAMP. No. 58.C.H., as fitted to No. 54.V.R. Dynamo Set, complete with best quality Bulb.

Diameter	- 4in.	Reflector	- Silver-plated.
Front Glass	- 3¼in.	Bulb -	2.5 volt, .3 amp. focal.

No. 58.C.H. Head Lamp only - Price **3/-**

Cable Code : 00574

BATTERY BOX. No. 61.G.H. Waterproof. Complete with metal fixing clips and switch.

No. 61.G.H. Battery Box only - Price **2/3**

Cable Code : 00575

CABLE. Best quality, silk covered, with terminals.

No. 666 Complete with Cable, Battery, and Bulb **5/6**

Cable Code : 00843

Replacement Battery No. B.23, **8d.** each.
Replacement Bulb No. 253, **5d.** each.

H. Miller & Co Ltd 1935 (10)

Sets for Cycles

THE Miller No. 666 was introduced to meet the demand for a lamp with a separate container — which may be attached to the top tube of the cycle and connected to the head lamp by cable. The round head lamp is particularly neat in appearance and is as fitted to our Dynamo Set (No. 54.V.R.). The battery container takes a standard 2-cell battery, is entirely weatherproof, and light in weight.

SPECIFICATION

HEAD LAMP. No. 58.C.H., as fitted to No. 54.V.R. Dynamo Set, complete with best quality Bulb.

Diameter	4in.	Reflector	Silver-plated.
Front Glass	3¼in.	Bulb	2.5 volt, .3 amp. focal.

No. 58.C.H. Head Lamp only - **Price 3/-**

Cable Code : 00574

BATTERY BOX. No. 61.G.H. Waterproof. Complete with metal fixing clips and switch.

No. 61.G.H. Battery Box only - **Price 2/3**

Cable Code : 00575

CABLE. Best quality, silk covered, with terminals.

No. 666 Complete with Cable, Battery, and Bulb **5/6**

Cable Code : 00643

Replacement Battery No. B.23, **8d.** each.
Replacement Bulb No. 253, **5d.** each.

The BROADBEAM

As fitted to Miller Lamps Nos. 57.V.R.,

THE OLD WAY

THIS startling new improvement in the construction of electric cycle lamps gives a beam practically double the normal width of that emitted by a lamp fitted with an ordinary reflector. Contrasting impressions, based on photographic records, of the vast improvement achieved are illustrated above. It will be seen that the central floodlight is by no means decreased in intensity. What happens is that an additional area is now illuminated — curbstones, hedges, pedestrians, etc., which previously had to be "guessed at," are now brought within the cyclist's

H. Miller & Co Ltd 1935 (11)

REFLECTOR

7.R.7, 6.R.6, 5.R.5, 4.R.4, 444

THE NEW WAY

view, so that greater speed and greater safety are assured.

The constructional details which make this improvement possible are clearly visible from the illustrations of all the lamps listed at the head of these pages and described on pages 52 to 61.

Cyclists should note that the Broad Beam Reflector is a Miller innovation and fitted exclusively to Miller Lamps.

H. Miller & Co Ltd 1935 (12)

Bulbs for Cycle Dynamo and Battery Lamps

Miller Bulbs for Cycle Dynamo have been selected after exhaustive tests. They are specially designed and constructed to withstand the varying output of a cycle dynamo set. They are definitely superior to the cheaper varieties on the market, and will outlast many bulbs of this type.

Miller Battery Lamp Bulbs have a focal spot, the filament having been designed to give a maximum concentration of light for a given wattage. They are packed in boxes as illustrated.

For Cycle Dynamo Lamp

SPECIFICATION.

No. D.B.4.	Head Lamp	...	4 volt .3 amp., Helix filament.
No. C.A.23.	Tail Lamp	...	5 volt .15 amp.

Supplied in boxes of 24 bulbs, **20/-** per box.

Cycle Dynamo Lamp Bulbs ... ... **10d.** each.

For Battery Lamps

SPECIFICATION.

No. 253.	2.5 volt .3 amp., Focal filament	...	... **6d.** each.

Supplied in boxes containing 24 bulbs ... **10/-** per box.

44

H. Miller & Co Ltd 1930

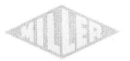

No. 324RC CYCLE DYNAMO SET

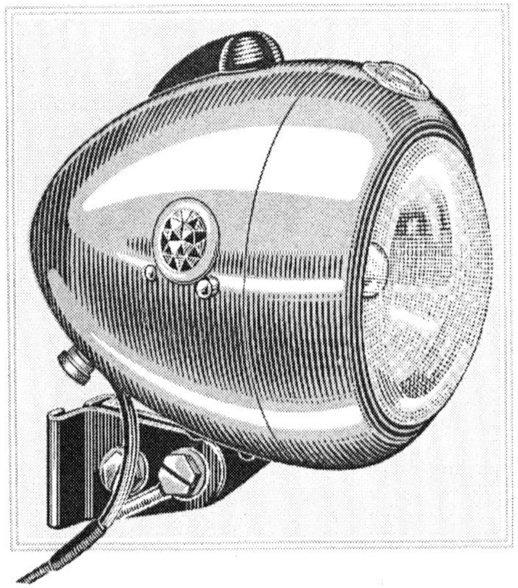

DYNAMO.

Combined Tail Lamp and approved prismatic reflector (N.P.L. Tested), incorporated in dynamo, making one compact unit and giving added protection to rider. Tail Lamp Bulb designed to resist road shocks and give longer life. New type totally-enclosed pivoting action ensuring greater efficiency with neater appearance. Mud-shield fitted as standard. **The lightest 6 watt dynamo yet produced.** This should appeal to Club riders. All-chrome finish.

Two-Year Guarantee

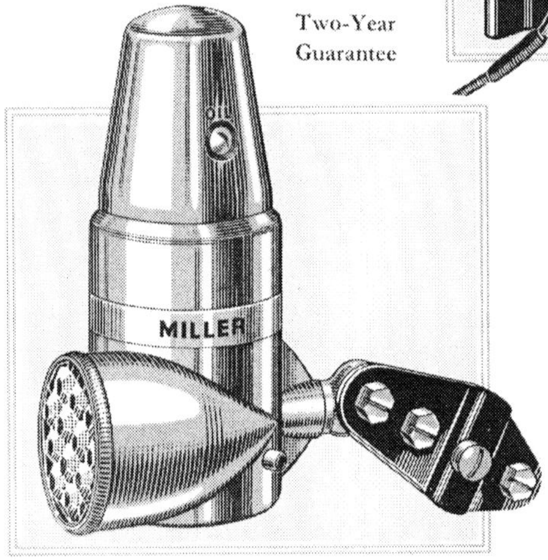

HEAD LAMP.

Improved streamlined, all - chrome finish brass Lamp, with no loose inside wires. Quick-release hinged front with moulded glass, giving greater security and eliminating breakage. Provision for stand-by battery. Car type main bulb.

S P E C I F I C A T I O N

Head Lamp No. 4.R.

Broadbeam Reflector	*3⅜in.*
Moulded Front Glass	*3⅝in.*
Main Bulb	*6 volt, 1.0 amp.*
Pilot Bulb	*4 volt, .3 amp.*
Three-way Switch.	
Weight	*15 ozs.*
Finish	*All Chrome*

Dynamo No. 32.C.2.

Capacity	*6 volt, 6 watt.*
Height	*4⅝in.*
Weight	*25½ ozs.*
Tail Lamp Bulb	*6 volt, 0.04 amp.*
Reflector Glass	*1⅛in. diameter.*
Clip	**Universal, to fit all types of stays.**
Finish	*All Chrome.*

No. 324RC Complete, including Cable and Clips **25/-**
Cable Code : 60818

Battery—No. B.34.		**5d.** each.
Replacement Bulbs.	Main—No. 610, 6 volt, 1.0 amp.	**1/3** each.
	Pilot—No. 453.F., 4 volt, .3 amp.	**4d.** each.
	Tail—No. 604, 6 volt, .04 amp.	**6d.** each.

Page 3

H. Miller & Co Ltd 1939 (1)

LIGHTING THE KING'S HIGHWAY THROUGH FIVE SUCCESSIVE REIGNS

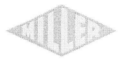

No. 324R CYCLE DYNAMO SET

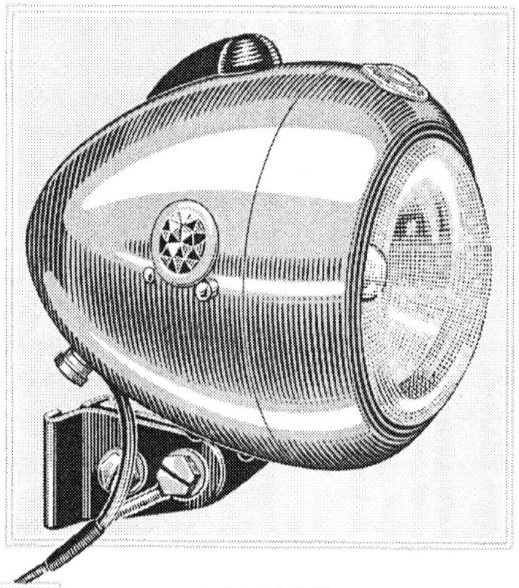

DYNAMO.

Combined Tail Lamp and approved prismatic Reflector (N.P.L. Tested) incorporated in dynamo, making one compact unit and giving added protection to rider. Tail Lamp Bulb designed to resist road shocks and give longer life. Totally-enclosed pivoting action, ensuring greater efficiency with neater appearance. Mud-shield fitted as standard. **The lightest 6 watt dynamo yet produced.** This should appeal to club riders.

Two-Year Guarantee

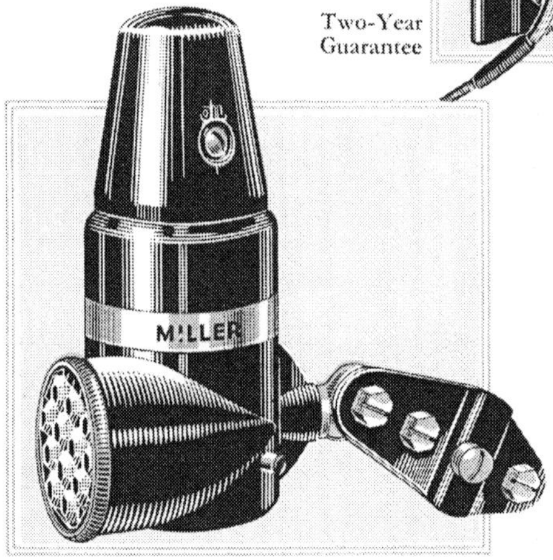

HEAD LAMP.

Improved streamlined, all - chrome finish brass Lamp with no loose inside wires. Quick-release hinged front with moulded glass, giving greater security and eliminating breakage. Provision for stand-by battery. Car type 6 watt gas-filled main bulb.

S P E C I F I C A T I O N

Head Lamp No. 4.R.

Broadbeam Reflector	*3⅜in.*
Moulded Front Glass	*3⅝in.*
Main Bulb (gas filled)	*6 volt, 1.0 amp.*
Pilot Bulb	*4 volt, .3 amp.*
Three-way Switch.	
Weight	*15 ozs.*
Finish	*All Chrome*

Dynamo No. 32.R.2.

Capacity	*6 volt, 6 watt.*
Height	*4⅝in.*
Weight	*25½ ozs.*
Tail Lamp Bulb	*6 volt, .04 amp.*
Reflector Glass	*1⅛in. diameter.*
Clip	**Universal, to fit all types of stays.**
Finish	*High lustre ebony, chromium parts.*

No. 324R Complete, with Cable and Clips **23/6**
Cable Code : 60793

Battery—No. B.34.		**5d.** each.
Replacement Bulbs.	Main—No. 610, 6 volt, 1.0 amp.	**1/3** each.
	Pilot—No. 453.F., 4 volt, .3 amp.	**4d.** each.
	Tail—No. 604, 6 volt, .04 amp.	**6d.** each.

Page 4

MILLER THE LAMPS THAT WON'T GO OUT

H. Miller & Co Ltd 1939 (2)

'LIGHTING THE KING'S HIGHWAY THROUGH FIVE SUCCESSIVE REIGNS'

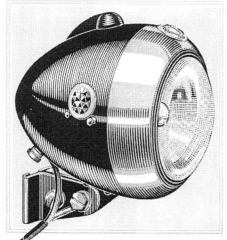

No. 325 CYCLE DYNAMO SET

DYNAMO.

Combined Tail Lamp and approved prismatic Reflector (N.P.L. Tested) incorporated in dynamo, making one compact unit and giving added protection to rider. Tail Lamp Bulb designed to resist road shocks and give longer life. New type totally-enclosed pivoting action, ensuring greater efficiency with neater appearance. Mudshield fitted as standard. **The lightest 6 watt dynamo yet produced.** This should appeal to club riders.

Two-Year
Guarantee

HEAD LAMP.

Redesigned streamlined lightweight model, specially treated to prevent rusting. Quick-release hinged front, chromium-plated brass. Car type main bulb and stand-by battery space. Finished with high-gloss scratchproof enamel.

S P E C I F I C A T I O N

Head Lamp No. 5.R.

Broadbeam Reflector	·	3¼in
Moulded Front Glass ·	·	3¼in.
Main Bulb ·	·	6 volt, 1.0 amp.
Pilot Bulb	·	4 volt, .3 amp.
Three-way Switch.		
Weight	·	13½ ozs.
Finish Chrome-plated brass rim, ebony body.		

Dynamo No. 32.R 2.

Capacity	·	6 volt, 6 watt.
Height	·	4½in.
Weight	·	25½ ozs.
Tail Lamp Bulb	·	6 volt, 0.04 amp.
Reflector Glass	·	1½in. diameter.
Clip - Universal, to suit all types of stays.		
Finish High lustre ebony, chromium parts.		

Page 5

No. 325	Complete, with Cable and Clips		21/-
	Cable Code : 00794		
Battery—No. B.34.		·	5d. each.
Replacement Bulbs.	Main—No. 619, 6 volt, 1.0 amp.	·	1/3 each.
	Pilot—No. 453.F. 4 volt, .3 amp.	·	4d. each.
	Tail—No. 604, 6 volt, 0.04 amp.	·	6d. each.

MILLER — THE LAMPS THAT WON'T GO OUT

H. Miller & Co Ltd 1939 (3)

LIGHTING THE KING'S HIGHWAY THROUGH FIVE SUCCESSIVE REIGNS'

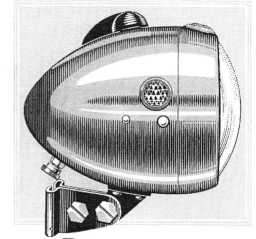

No. 537 & 557T CYCLE DYNAMO SETS

DYNAMO.

This 15½ oz. Dynamo weighs only 4.75 ozs. per watt ! It can be adapted for either front or rear wheel drive and used with or without tail lamp. Half chrome finish.

HEAD LAMP.

Modern lightweight, streamlined, car type with press-button front opening. Provision for stand-by battery. Car type main bulb, gas filled. All chrome finish.

Two-Year
Guarantee

TAIL LAMP.

Fitted with approved prismatic reflector (N.P.L. Tested) and bulb specially designed to withstand road shocks and give longer life.

S P E C I F I C A T I O N

Head Lamp No. 7.

Broadbeam Reflector	·	3¼in. diameter.
Glass	·	3¼in. diameter.
Main Bulb	·	6 volt, .5 amp.
Pilot Bulb	·	4 volt, .3 amp.
Three-way Switch.		
Weight	·	12 ozs.
Finish	·	All chrome.

Dynamo No. 53.

Capacity	·	6 volt, 3.24 watt.
Height	·	4in.
Weight	·	15½ ozs.
Clip - Universal, to suit all types of stays.		
Finish	·	Half-chrome.

Tail Lamp No. 597.

Reflector Glass	·	H.P.L. 1½in. diameter.
Bulb	·	6 volt, .04 amp.
Finish	·	High lustre ebony.

Replacement Bulbs.

Main—No. 605, 6v., .5 amp.	·	1/- each.
Pilot—No. 453.F., 4v., .3 amp.	·	4d. each.
Tail—No. 604, 6v. .04 amp.	·	6d. each.

No. 557T	Set, with Tail Lamp	20/-
	Cable Code : 00929	
No. 557	Set, less Tail Lamp	18/6
	Cable Code : 00930	
Battery—No. B.34	· · · · · ·	5d.

Page 6

MILLER THE LAMPS THAT WON'T GO OUT

H. Miller & Co Ltd 1939 (4)

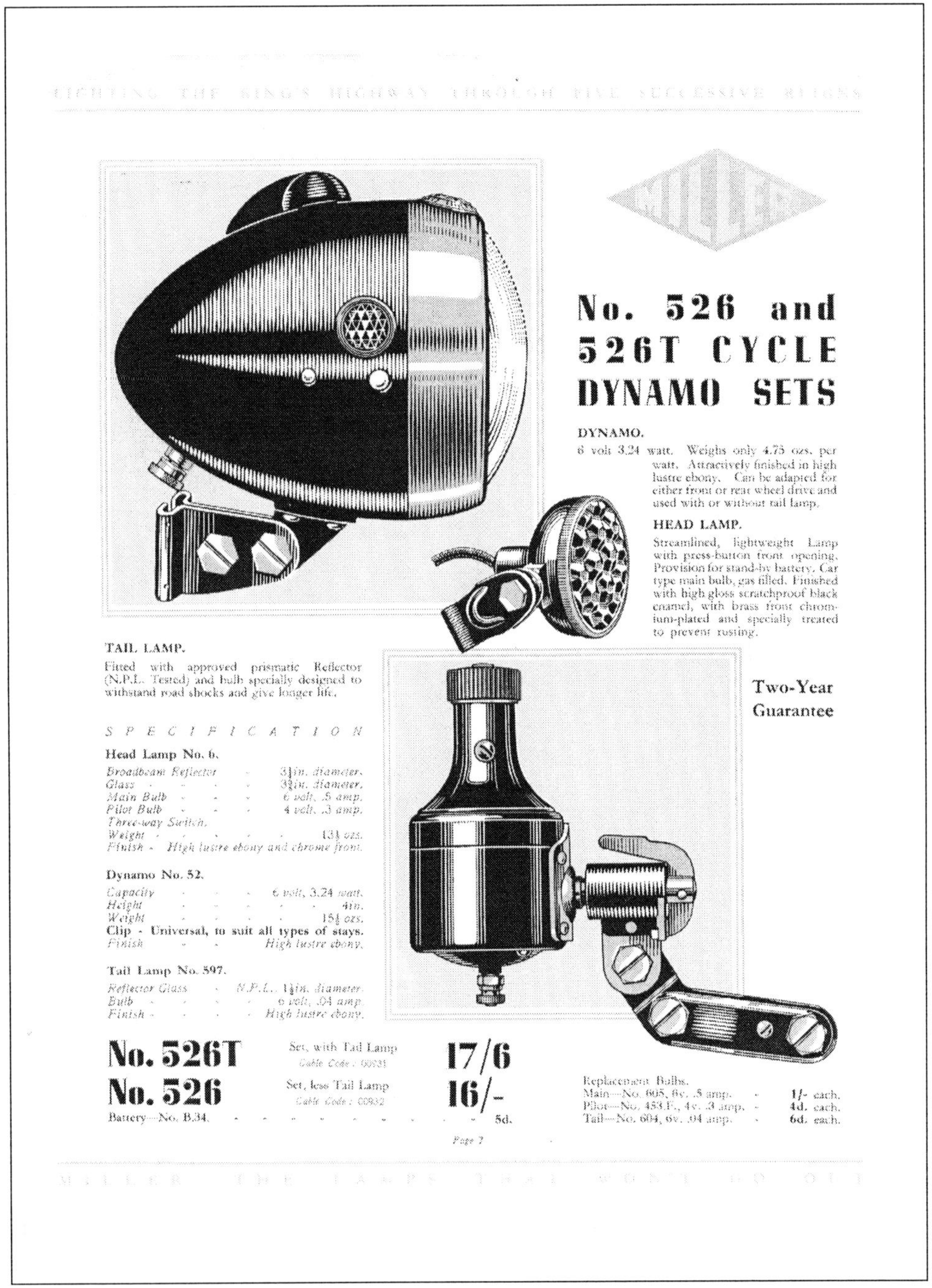
LIGHTING THE KING'S HIGHWAY THROUGH FIVE SUCCESSIVE REIGNS
MILLER
No. 526 and 526T CYCLE DYNAMO SETS
DYNAMO.
6 volt 3.24 watt. Weighs only 4.75 ozs. per watt. Attractively finished in high lustre ebony. Can be adapted for either front or rear wheel drive and used with or without tail lamp.
HEAD LAMP.
Streamlined, lightweight Lamp with press-button front opening. Provision for stand-by battery. Car type main bulb, gas filled. Finished with high gloss scratchproof black enamel, with brass front chromium-plated and specially treated to prevent rusting.
TAIL LAMP.
Fitted with approved prismatic Reflector (N.P.L. Tested) and bulb specially designed to withstand road shocks and give longer life.
Two-Year Guarantee
SPECIFICATION
Head Lamp No. 6.
Broadbeam Reflector - 3¼in. diameter.
Glass - - - - 3⅜in. diameter.
Main Bulb - - - 6 volt, .5 amp.
Pilot Bulb - - - 4 volt, .3 amp.
Three-way Switch.
Weight - - - - - 13½ ozs.
Finish - High lustre ebony and chrome front.
Dynamo No. 52.
Capacity - - - 6 volt, 3.24 watt.
Height - - - - - 4in.
Weight - - - - - 15¼ ozs.
Clip - Universal, to suit all types of stays.
Finish - - High lustre ebony.
Tail Lamp No. 597.
Reflector Glass - N.P.L. 1⅛in. diameter.
Bulb - - - - 6 volt, .04 amp.
Finish - - - - High lustre ebony.
No. 526T Set, with Tail Lamp Cable Code: 00931 17/6
No. 526 Set, less Tail Lamp Cable Code: 00932 16/-
Battery—No. B.34. - - - - - - - - - - 5d.
Replacement Bulbs.
Main—No. 605, 6v. .5 amp. - 1/- each.
Pilot—No. 453.F., 4v. .3 amp. - 4d. each.
Tail—No. 604, 6v. .04 amp. - 6d. each.
Page 7
MILLER THE LAMPS THAT WON'T GO OUT

H. Miller & Co Ltd 1939 (5)

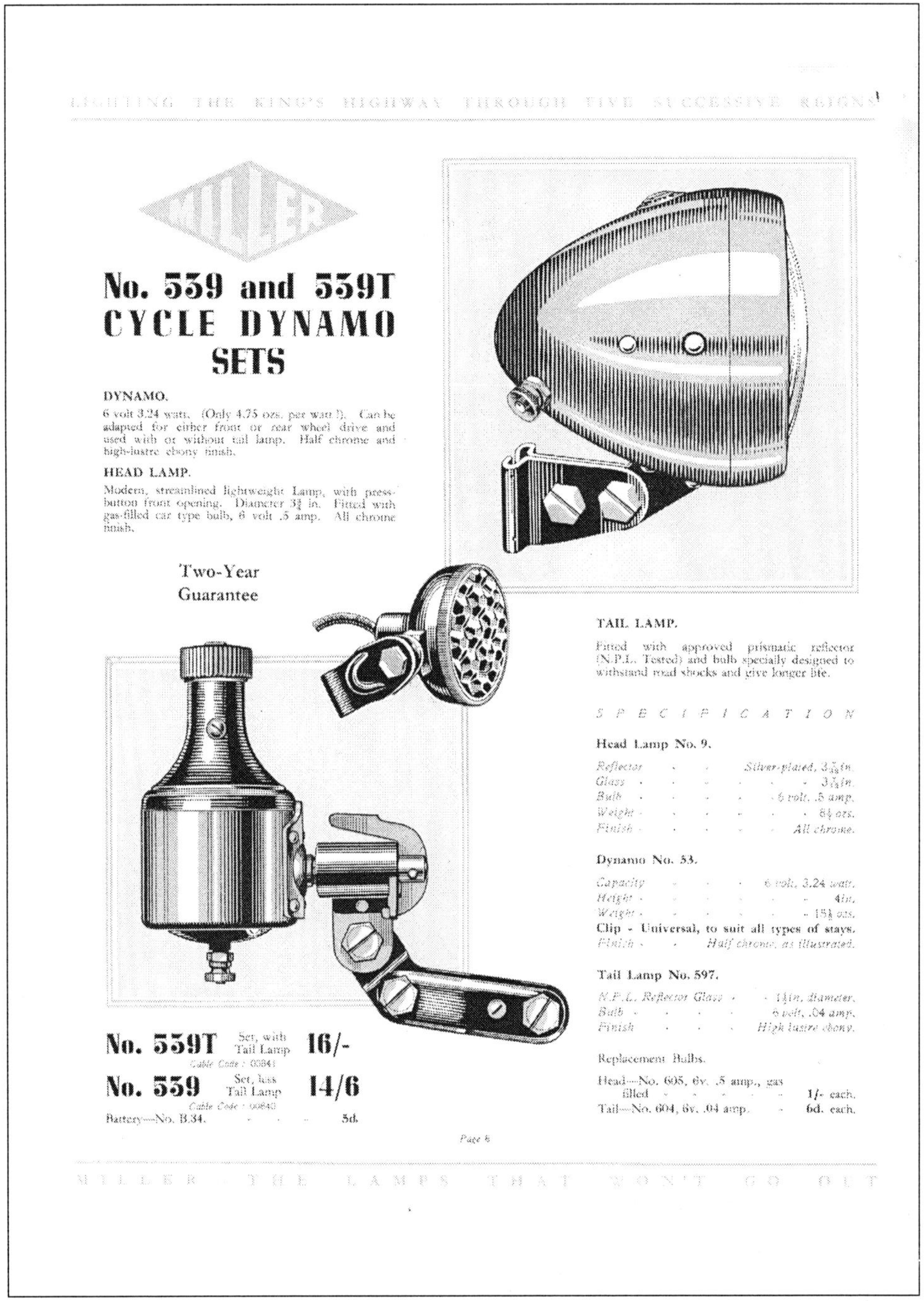
LIGHTING THE KING'S HIGHWAY THROUGH FIVE SUCCESSIVE REIGNS
MILLER
No. 539 and 539T CYCLE DYNAMO SETS
DYNAMO.
6 volt 3.24 watt. (Only 4.75 ozs. per watt!). Can be adapted for either front or rear wheel drive and used with or without tail lamp. Half chrome and high-lustre ebony finish.
HEAD LAMP.
Modern, streamlined lightweight Lamp, with press-button front opening. Diameter 3⅝ in. Fitted with gas-filled car type bulb, 6 volt .5 amp. All chrome finish.
Two-Year Guarantee
TAIL LAMP.
Fitted with approved prismatic reflector (N.P.L. Tested) and bulb specially designed to withstand road shocks and give longer life.
SPECIFICATION
Head Lamp No. 9.
Reflector - - Silver-plated, 3 7/16in.
Glass - - - - - - 3 7/16in.
Bulb - - - - - 6 volt, .5 amp.
Weight - - - - - - 8½ ozs.
Finish - - - - - All chrome.
Dynamo No. 53.
Capacity - - - 6 volt, 3.24 watt.
Height - - - - - - 4in.
Weight - - - - - - 15¼ ozs.
Clip - Universal, to suit all types of stays.
Finish - - Half chrome, as illustrated.
Tail Lamp No. 597.
N.P.L. Reflector Glass - - 1⅛in. diameter.
Bulb - - - - 6 volt, .04 amp.
Finish - - - - High lustre ebony.
No. 539T Set, with Tail Lamp Cable Code: 00841 16/-
No. 539 Set, less Tail Lamp Cable Code: 00840 14/6
Battery—No. B.34. - - - 5d.
Replacement Bulbs.
Head—No. 605, 6v. .5 amp., gas filled - - - - - 1/- each.
Tail—No. 604, 6v. .04 amp. - 6d. each.
Page 6
MILLER THE LAMPS THAT WON'T GO OUT

H. Miller & Co Ltd 1939 (6)

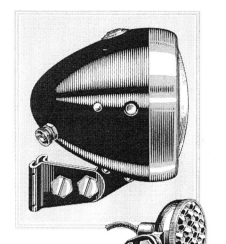

No. 528 and 528T CYCLE DYNAMO SETS

DYNAMO.

Ultra-light dynamo — 4.75 oz. per watt. Adaptable for front or rear wheel fitting and can be used with or without tail lamp. Attractively finished in high lustre ebony with chromium-plated parts.

HEAD LAMP.

Lightweight, streamlined car type Lamp with press-button front opening. Finished with high-gloss scratchproof black enamel, with chromium-plated brass front, specially treated to prevent rusting. Standard 6 volt .5 amp. bulb, gas filled.

Two-Year Guarantee

TAIL LAMP.

Fitted with approved prismatic reflector (N.P.L. Tested) and bulb specially designed to withstand road shocks and give longer life.

S P E C I F I C A T I O N

Head Lamp No. 8.

Reflector	-	Silver-plated, 3½ in.
Glass	-	3½ in.
Bulb	-	6 volt, .5 amp., gas filled.
Weight	-	6½ ozs.
Finish	-	High-gloss scratchproof enamel, with chrome-plated brass front.

Dynamo No. 52.

Capacity	-	6 volt, 3.24 watt.
Height	-	4 in.
Weight	-	15½ oz.
Clip - Universal, to suit all types of stays.
Finish - High lustre ebony, with chrome parts.

Tail Lamp No. 597.

N.P.L. Reflector Glass	-	1½ in. diameter.
Bulb	-	6 volt, .04 amp.
Finish	-	High lustre ebony.

No. 528T Set, with Tail Lamp **14/-**
Cable Code : 00829

No. 528 Set, less Tail Lamp **12/6**
Cable Code : 00828

Replacement Bulbs.
Head—No. 605, 6v. .5 amp. gas-filled 1/- each.
Tail—No. 604, 6v. .04 amp. - 6d. each.

Page 9

H. Miller & Co Ltd 1939 (7)

CYCLE DYNAMO SET COMPONENTS

No. 4.R. HEAD LAMP, as supplied with sets Nos. 324.R.C. and 324.R. Fitted with 6 volt 1.0 amp. gas-filled bulb.
Price 11/-, complete with cables.
Cable Code : 00798.

No. 8.R. HEAD LAMP, as supplied with set No. 325. Fitted with 6 volt 1.0 amp. gas-filled bulb.
Price 8/6, complete with cables.
Cable Code : 00799.

No. 7 HEAD LAMP, as supplied with sets Nos. 537 and 537.T. Fitted with 6 volt .5 amp. gas-filled bulb.
Price 10/-, complete with cables.
Cable Code : 00928.

No. 9 HEAD LAMP, as supplied with sets Nos. 539 and 539.T. Fitted with 6 volt .5 amp. gas-filled bulb.
Price 6/-, complete with cables.
Cable Code : 00844.

No. 8 HEAD LAMP, as supplied with sets Nos. 528 and 528.T. Fitted with 6 volt .5 amp. gas-filled bulb.
Price 4/6, complete with cables.
Cable Code : 00842.

No. 6 HEAD LAMP, as supplied with sets Nos. 528 and 528T. Fitted with 6 volt .5 amp. gas-filled bulb.
Price 8/-, complete with cables.
Cable Code : 00871.

No. 32.C.2. DYNAMO, as supplied with set No. 324.R.C. 6 volt, 6 watt capacity.
Price 14/6 each.
Cable Code : 00820.

No. 32.R.2. DYNAMO, as supplied with sets Nos. 324.R. and 325. 6 volt, 6 watt capacity.
Price 13/- each.
Cable Code : 00851.

No. 53 DYNAMO, as supplied with sets Nos. 537, 537.T., 539, and 539.T. 6 volt, 3.24 watt capacity.
Price 9/- each.
Cable Code : 00879.

No. 52 DYNAMO, as supplied with sets Nos. 528, 528.T., 528, and 528.T. 6 volt, 3.24 watt capacity.
Price 8/6 each.
Cable Code : 00843.

No. 597 TAIL LAMP, as supplied with sets Nos. 537.T., 539.T., 528.T. and 528.T. Fitted with 6 volt, .04 amp. bulb.
Price 1/6 each.
Cable Code : 00847.

No. 27 DYNAMO MUDSHIELD. To fit most makes of cycle dynamos.
Price 6d. each.
Cable Code : 00806.

No. 4.R. Head Lamp

No. 7 Head Lamp

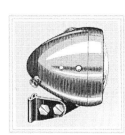

No. 9 Head Lamp

No. 597 Tail Lamp

No. 32.C.2. Dynamo

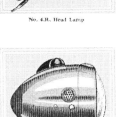

No. 53 Dynamo

No. 27 Dynamo Mudshield

Page 10

H. Miller & Co Ltd 1939 (8)

Miller Battery

AN entirely new MILLER line is this Double Battery Lamp, designed to take two flash lamp batteries. This enables an 8 volt lamp bulb to be used, which results in a very intense beam of light for dimly lighted streets and lanes. Current wastage, however, in brilliantly lit streets is avoided by use of the dimmer switch, conserving the current yet giving reasonably good illumination. This feature intelligently used with new batteries will prevent bulb breakage from overload.

SPECIFICATION

Diameter - - - - 4½in.	Switch - - - Three-way.
Diameter of Front - - 4in.	Brackets - - - Side Screw.
Diameter of Front Glass - 3¹¹⁄₁₆in.	Weight (without Battery) - 15 oz.
Bulb - - - 8v. .1a.	Finish - Ebony, with chromium-
Reflector, BROADBEAM, nickel-plated.	plated parts.

No. 8.R.8   Price, complete with Miller 8 volt Bulb and two Miller Batteries   **7/-**

Cable Code : 00882

No. B.34.	Replacement Batteries	-	**10d.** per pair.
No. 851.	Replacement Bulbs	- -	**5d.** each.

H. Miller & Co Ltd 1939 (9)

 # No. 5.A.5 BATTERY LAMP

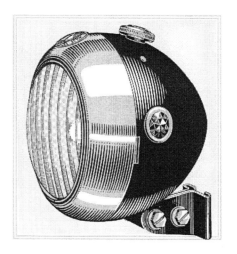

A NEW streamlined model. Rustproofed and finished with high-gloss scratchproof black enamel. Quick-release hinged front, chromium-plated brass. Silver-plated broad beam reflector. Screw-down "on" and "off" switch.

SPECIFICATION

Diameter - - - - 4½in.
Ribbed Glass - - 3½in.
Bulb, Focal, 2.5 volt, .3 amp.
Reflector, Silver-plated, 3½in.
Switch, Screw-down, "on" and "off."
Weight (less battery) 12 ozs.

No. 5.A.5   Complete with Battery and Bulb   *Cable Code : G0723*   **5/-**

Ivory Finish - - - **3d.** extra.

Replacement Battery No. B.23	- - -	**8d.** each.
Replacement Bulb No. 325	- - -	**3d.** each.

Page 12

H. Miller & Co Ltd 1939 (10)

LIGHTING THE KING'S HIGHWAY THROUGH FIVE SUCCESSIVE REIGNS

No. 4.A.4 BATTERY LAMP

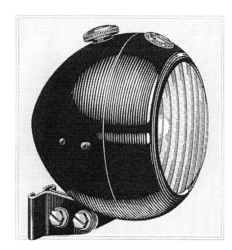

A POPULAR priced streamlined Head Lamp fitted with ribbed glass. Finished in high gloss scratchproof black enamel.

SPECIFICATION

Diameter of Front	4¼in.
Ribbed Glass	
Bulb, Focal, 2.5 volt, .3 amp.	
Reflector	3¼in.
Switch, Screw-down,	
	"on" and "off."
Weight (less battery)	12 ozs.

No. 4.A.4 Complete with Battery and Bulb **5/11**
Cable Code : 66621

Replacement Battery No. B.23 . . . . **8d.** each.
Replacement Bulb No. 253 . . . . **3d.** each.

Page 13

MILLER THE LAMPS THAT WON'T GO OUT

No. 252 BATTERY LAMP

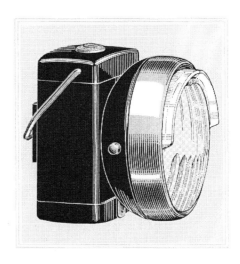

A REDESIGNED general purpose lamp. Substantially made and specially treated to prevent rusting. It has the advantage of a pilot bulb of low amperage which effects considerable saving of the battery when used in well-lighted streets and also acts as a reserve bulb. Moulded broadbeam lens giving side warning.

SPECIFICATION

Height	4in.
Front Glass	Moulded broadbeam.
Bulb:—	
Main	2.5 volt, .3 amp.
Pilot	2.5 volt, .06 amp.
Reflector	Silver-plated.
Finish	High lustre ebony with chromium-plated brass rim.

No. 252 Complete with Battery and Bulb **4/6**
Cable Code : 00793

Ivory Finish . . . **3d.** extra.

Replacement Battery, No. B.23 . . . . **8d.** each.
Replacement Bulbs: Main—No. 253, 2.5v., .3 amp. · **3d.** each.
Pilot—No. 251, 2.5v., .06 amp. · **3d.** each.

Page 14

MILLER THE LAMPS THAT WON'T GO OUT

H. Miller & Co Ltd 1939 (11)

H. Miller & Co Ltd 1939 (12)

CATALOGUES & ADVERTISMENTS

No. 242 BATTERY LAMP

AN interesting new model, a special feature of which is the large chromium plated front rim fitted with ribbed glass, giving a broad beam. Screw-down "On—off" switch. Finished with high-gloss scratchproof enamel specially treated to prevent rusting.

SPECIFICATION

Height · · · 4in.
Diameter of Front · 3in.
Glass · · · Ribbed.
Bulb 2.5 volt. .3 amp. focal.
Reflector · Silver-plated.
Weight (less battery) ·
Finish · High lustre ebony with chromium-plated brass rim.

No. 242 Complete with Battery and Bulb *Cable Code : 05791* **5/6**

Ivory Finish · · · **3d.** extra.

Replacement Battery, No. B.23 · · **8d.** each.
Replacement Bulbs. Main—No. 253, 2.5v., .3 amp. · **3d.** each.
Pilot—No. 251, 2.5v., .06 amp. · **3d.** each.

H. Miller & Co Ltd 1939 (13)

No. 3.L.5 BATTERY LAMP

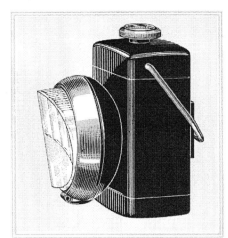

SQUARE pattern Lamp. Specially treated to prevent rusting. New design chromium-plated brass front rim with hasp fastening. Special lens to give long and broad beam, with side warning. Screw-down "on" and "off" switch. Stormproof. Finished with high-gloss scratchproof black enamel.

SPECIFICATION

Height · · 4in.   Weight (less Battery) 6½ ozs.
Lens · · 2⅛in.   Finish · High lustre ebony
Bulb Focal, 2.5 volt,   with chromium-
.3 amp.   plated brass rim.

No. 3.L.5 Complete with Battery & Bulb *Cable Code : 06683* **3/-**

Replacement Battery No. B.23 · · · **8d.** each.
Replacement Bulb (Main) No. 2503, 2.5 volt, .3 amp. · **3d.** each.

No. 222 BATTERY LAMP

SQUARE pattern Lamp. Specially treated to prevent rusting. New design chromium-plated brass front rim with hasp fastening. Screw-down "on" and "off" switch. Stormproof. Finished with high-gloss scratchproof black enamel.

SPECIFICATION

Height · · 4in.   Weight (less Battery) 6 ozs.
Glass · · 2⅛in.   Finish - High lustre ebony
Bulb Focal, 2.5 volt,   with chromium-
.3 amp.   plated brass rim.

No. 222 Complete with Battery & Bulb *Cable Code : 06869* **2/9**

Replacement Battery No. B.23 · · · **8d.** each.
Replacement Bulb No. 2503, 2.5 volt, .3 amp. · **3d.** each.

H. Miller & Co Ltd 1939 (14)

Set No. 536 T

The headlamp of this Set (3¾" dia. x 5¼") is finished in either Silver Grey or Black with chromium rim. Fitted with two 6 volt 0.5 amp. bulbs and with accommodation for stand-by battery. The lightweight 6 volt 3.24 watt capacity Dynamo combined with new style tail lamp is finished throughout in chromium plate.

Retail Price 40/-

The trouble-free sets for trouble-free SALES !

Set No. 535 T

The ultra lightweight headlamp of this set measures 2½" dia. x 3¾" and weighs only 5½ ozs. Finished throughout in lasting chromium plate. Dynamo as described above.

Retail Price 35/-

MILLER

— here are two winners always in popular demand.

No. 820

Chromium plated 3" bell metal brass dome. Smooth side-to-side action with five - point striker. Special non-slip band clip fits any type of handlebar.

Retail Price 6/-

No. 520

Rotary type bell with chromium plated 2⅜" bell metal brass dome. Also fitted with special non-slip band clip.

Retail Price 3/9

The range of Miller bells is featured on page 8.

H. MILLER & CO. LTD., ASTON BROOK STREET, BIRMINGHAM 6

H. Miller & Co Catalogue cover

LIGHTING THE KING'S HIGHWAY THROUGH FIVE SUCCESSIVE REIGNS

MILLER

No. 121 BATTERY LAMP

POPULAR priced square pattern Lamp. Specially treated to prevent rusting. Screwed chromium-plated or black front. Screw-down "on" and "off" switch. Stormproof. Finished with high-gloss scratchproof black enamel.

SPECIFICATION

Height	*4in.*
Glass	*2½in.*
Bulb, Focal,	*2.5 volt, .2 amp.*
Weight (less battery)	*6 ozs.*
Finish	*High lustre ebony with chromium-plated front.*

No. 121   Complete with Battery and Bulb   **2/6**
Cable Code : OGB19

Replacement Battery	No. B.23	· · · ·	8d. each.
Replacement Bulb	No. 2503	· ·	3d. each.

Page 17

MILLER · THE LAMPS THAT WON'T GO OUT

H. Miller & Co Ltd 1939 (15)

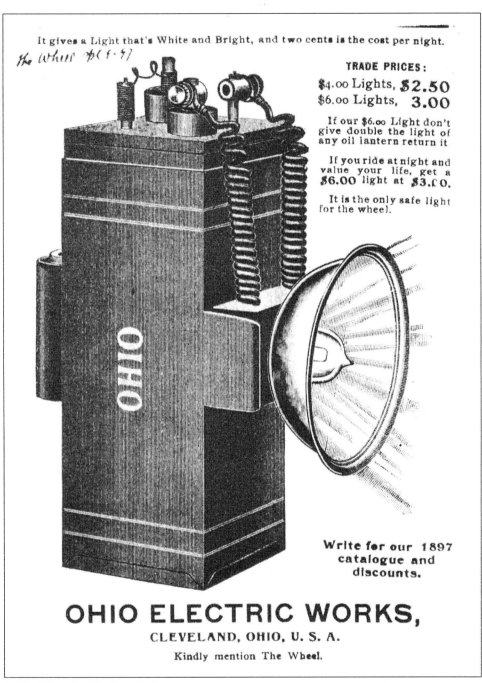

It gives a Light that's White and Bright, and two cents is the cost per night.

TRADE PRICES:

$4.00 Lights, **$2.50**

$6.00 Lights, 3.00

If our $6.00 Light don't give double the light of any oil lantern return it

If you ride at night and value your life, get a **$6.00** light at **$3.00.**

It is the only safe light for the wheel.

Write for our 1897 catalogue and discounts.

OHIO ELECTRIC WORKS,

CLEVELAND, OHIO, U. S. A.

Kindly mention The Wheel.

Ohio Electric Works 1897

FLASHLIGHT AND HEADLIGHT BRACKETS

FLASHLIGHT BRACKET

A popular bracket at a nominal price. Cadmium plated. Handle bar binder bolt fastening .... Stock No. 339

FLASHLIGHT BRACKET

This bracket fills a long felt demand for riders who desire a prompt adjustment of a flashlight in any direction Constructed of heavy steel, nickel plated .... Stock No. 342

FLASHLIGHT BRACKET

A good bracket at a nominal price. Fastens to handlebar. Satin plated. Stock No. 342A

HEAD LAMP BRACKET

Reversible right and left—adjustable —nickel plated .... Stock No. 1012

FLASHLIGHT BRACKET

An ideal bracket for making an inexpensive and attractive lamp out of a tubular flashlight. Nickel plated. Stock No. 340

FLASHLIGHT BRACKET

Double spring clip flashlight bracket. Satin plated. Furnished complete with bolts and nuts for attaching to handle bar .... Stock No. 338

HEAD LAMP BRACKET

Nickel plated—Lined with felt. Made to fit 1¼, 1⅜, 1½ and 1⅝ inch head. Stock No. 1013

FLASHLIGHT BRACKET

With the adjustable features incorporated in this bracket, it will accommodate different diameter flashlights. Made of heavy steel, nickel plated. Stock No. 341

FLASHLIGHT BRACKET

Quickly and easily attached to any handle bar. Made with rubber rollers which act as cushions and absolutely prevent rattling and marring of flashlight case. Easily adjusted to any angle desired. Black crystal lacquer finish .... Stock No. 338A

HEAD LAMP BRACKET

Adjustable, nickel plated—reversible right and left .... Stock No. 1014

Oil and Cycle Supply Company USA 1939

CYCLE DYNAMO LIGHTING SET.

No. 33.

This Set has been designed for the rider who requires trouble-free lighting without the use of oil or carbide.

A very neat and compact dynamo supplies the current to a large front lamp and small rear lamp.

The concentrated beam of light from this lamp is such that an excellent riding light is obtained, under all conditions.

The Head Lamp No. 34 is fitted with a change over switch and provision is made inside the lamp for carrying a 3.5 volt ordinary flashlamp battery; by means of the switch the rider can change over from the dynamo to the battery when the machine is stationary, i.e., in traffic holdups.

The dynamo supplies ample current for both lamps even when the cycle is propelled at only walking pace.

The Set is supplied complete with all cable, bulbs, etc., and is ready for immediate attachment to the machine.

PRICE.

Complete with Tail Lamp -   £1   1   0 Set.
Less Tail Lamp   -   -   19   6 "

No. 34 Headlamp, with change over switch to take 3.5 flash lamp battery. Diameter of front, 4-in. Complete with bulb, *less* battery.

PRICE, 4/-

No. 32 Tail Lamp, fitted with adjustable bracket and a 1⅜-in. ruby glass. **PRICE, 1/6**

CYCLE DYNAMO LIGHTING SET.

This Set Comprises:

No. 34 Headlamp.

Dynamo.

No. 32 Tail Lamp.

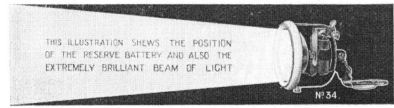

THIS ILLUSTRATION SHEWS THE POSITION OF THE RESERVE BATTERY AND ALSO THE EXTREMELY BRILLIANT BEAM OF LIGHT

Powell & Hanmer 1929 (1)

ELECTRIC CYCLE LAMPS.

P. & H. ELECTRIC DRY BATTERY CYCLE LAMPS.

No. 30. MOONLITE.

This Lamp has been designed to accommodate a 2 Cell Dry Battery. Manufactured on sound principles; beautifully finished, strongly made, it will withstand very rough usage.

The means for attachment to the Cycle is provided for by a bracket rivetted to the back of the lamp and is securely locked to the lamp bracket by means of a coin-slotted screw. When screwed up tightly, the Lamp is very firm on the bracket and will not become loose with vibration.

A positive Switch is contained in the Lamp being worked by a Push Button over the top; when being used as a Hand or Flash Lamp, contact is made by a slight pressure on this Button, which, when released, immediately switches off the current.

For use as a Cycle Lamp press the Switch down and turn right for locking in the "on" posistion.

The Handle which is detachable is held firmly in position by a spring catch at the back of the lamp when used on a cycle. To release the top cap of lamp the spring handle must be pulled off first.

The Ebony finished body with heavily nickel-plated front gives the lamp a very neat and pleasing appearance.

Height, 3⅞-ins. Front, Nickel Plated, 2⅜-ins. Reflector, Silver-Plated, 2⅜-ins.

Including Battery and No. 25 2·5 **Bulb.**

PRICE, 3/6 complete.

No. 31. EASYLITE.

Is designed for use with a 2 Torch Unit Dry Cell Batteries thereby giving long life before replenishing.

As with our No. 30, this model is also fitted with a detachable handle, a positive switch and a locking screw for use as a Cycle Lamp.

Bolder in appearance than our No. 30, and opening at the front instead of the top for renewing the Batteries.

At the back of the Reflector of this lamp a spare Bulb Holder is fitted. The body of this lamp is finished in ebony, with heavy nickel-plated front.

Height, 3⅞-ins. Front, nickel-plated, 2⅜-ins. Reflector, silver-plated, 2½-ins.

Including 2 Torch Unit Dry Cell Batteries and No. 25 2·5 **Bulb.**

PRICE, 5/- complete.

MOONLITE.

No. 30.

SPARE BATTERY,
PH No. 800. **1/-** each.

SPARE BULBS.

No. 25. 2·5 **5d.** each.

EASYLITE.

No. 31.

TORCH UNIT
DRY CELL BATTERY.

2 cells required **4½d.** each.

SPARE BULBS.

No. 25. 2·5 **5d.** each.

The above can be fitted with Red Glasses
for use as Rear Lamps.

Powell & Hanmer 1929 (2)

N° 35 S.B 6'- Complete   N° 35 5'6 Complete

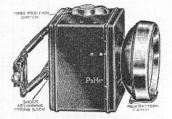

No. 35 S.B.
With Spring Back and Nickel-Plated Fittings.
Price, 6/- complete.

No. 35 C.P.
With Spring Back and Chromium-Plated Fittings.
Price, 6/6 complete.

This lamp has been designed to meet the needs of the rider who wishes to "Fit and Forget" for a season, and has proved itself an extremely popular model.

Fitted with specially designed Spring Back, incorporating shock absorbing device.

As will be seen, this shows alternative fittings for the No. 35 Lamp. This bracket has been designed to fit on the handlebar stem, and the lamp can be fitted in this manner without fouling the brake rods.

Provision is made for carrying two No. 800 Batteries, with 3-position Switch, to provide a full light (5.5 volt) or a dim light for use in Town when desired.

It is fitted with a Sterling Silvered-Plated Reflector. **All-Brass Front**, and gives a wonderful beam of light.

Diameter of Front, 3½".

No. 35. Finish, Ebony and Nickel-Plated   ...   5/6

GIVES FIFTY HOURS' CONTINUOUS LIGHT

No. 30, as illustrated, Ebony and Nickel-Plated   3/-
No. 30D, with 3-position Dimmer Switch   ...   3/6

This is a neat, well-made lamp, and one which has given satisfaction for many years.

Fitted with Sterling Silver-Plated Reflector, and All-Brass Front Nickel-Plated.

No. 30D is fitted with a 3-position dimmer switch so that bright or dim light may be used as desired.

Takes No. 800 Battery and 2.5 v. Bulb.

This improved model represents the finest value offered in Cycle Battery Lamps.

Made throughout of heavy gauge material, and embodies many new features.

The front of the lamp opens to give easy access to the battery, and is entirely weatherproof.

A screw-down Switch (cut thread) is fitted, and a highly polished Reflector gives an excellent beam of light. Is fitted with a handle which can be securely locked at back when lamp is used on cycle.

Takes the No. 800 Battery and 2.5 v. Bulb.
Finish, All Ebony.

N° 30 3'- Complete   N° 15 2'6 Complete

N° 29 4'- Complete   N° 31 4'1 Complete

A well-designed and thoroughly efficient lamp, the front of which opens for easy access to battery. The lamp is watertight, and has a Sterling Silver-Plated Reflector. Screw-down switch (cut thread). Can be used as Hand Inspection or Cycle Lamp. When used as a cycle lamp the handle is held in position by means of a spring clip at the back of lamp.

A new feature regarding this lamp is that it is fitted with a newly designed All-Brass Front, with a fold-over catch, which is simple in operation and very secure.

Diameter of Front, 3½".

Takes No. 800 Battery and 2.5 v. Bulb.

No. 29. Ebony and Nickel-Plated   4/- complete.
No. 29. Ebony and Chrome   ...   4/3 complete.

This lamp is designed to take two P. & H. U.2 Batteries.

The lamp opens at the front to give easy access to batteries, and being fitted with a bold All-Brass Front, heavily Nickel-Plated, gives the lamp a good appearance.

Provision is made at the back of the Reflector for carrying a spare bulb. Fitted with screw-down Switch, Brass Reflector Sterling Silver-Plated.

Diameter of Front, 2½".

Takes two U.2 Batteries and 2.5 v. Bulb.
Finish, Ebony and Nickel-Plated.

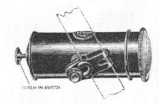

TAIL LAMP

This lamp has been designed to meet the demands for a Battery Tail Lamp, and is proving itself to be a popular model with cyclists.

Is fitted with a positive screw-in type Switch and Bayonet Front, which securely fastens same to body of lamp.

Fitted with 1½" Ruby Bi-Convex Lens.

Has an adjustable clip which can be supplied for fitting to either D stay or round forks.

Takes U. 2 Battery and 2.5 v. Bulb.
Finish, All Ebony.

No. 39 Battery Box Set   ...   ...   5/6 complete.

Those who desire this form of lighting will find this an exceptionally reliable set.

The Head Lamp is of pleasing design, having a 4" Front and Sterling Silver-Plated Reflector.

The Battery Box is fitted with all-metal clips and spring grip terminals.

A screw-in type Switch is used, ensuring perfect contact under all conditions.

Takes P. & H. No. 800 Battery and 2.5 v. Bulb.
Finish, All Ebony.

No. 27 Set   ...   .   7/3 complete.

As above, but with torpedo-shaped Head Lamp with adjustable bracket.

N° 54 2'- Complete   N° 39 5'6 Complete

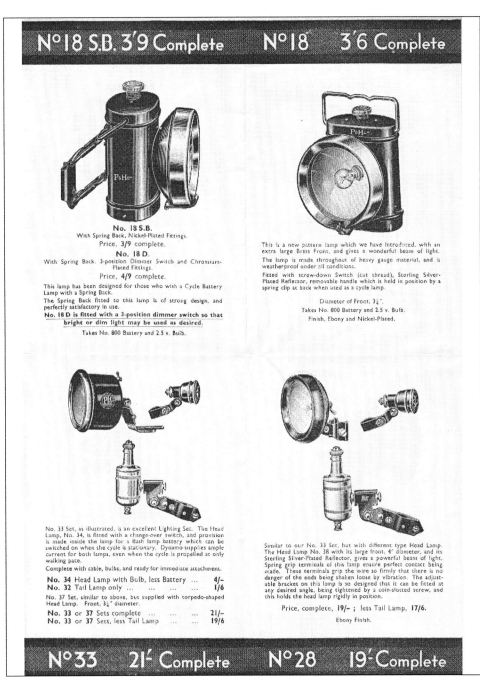

Powell & Hanmer 1931 (3)

Powell & Hanmer 1935

CYCLE BATTERY LAMPS

are fitted with specially designed spiral filament bulbs to withstand vibration and high voltage of new batteries.

No. 35. Stem Fitting, Ebony and N.P. Price **5/6** complete.

No. 35 S.B. With Spring Back and Nickel Plated Fittings. Price **6/-** complete.

No. 35 C.P. With Spring Back and Chromium Plated Fittings. Price **6/6** complete.

No. 35 is designed to fit on the handle-bar stem without fouling the brake rods. Nos. 35 S.B. and C.P. are fitted with specially designed spring backs incorporating shock absorbing device. Provision is made for carrying 2— No. 800 Batteries, with 3-position switch, giving a full light (5.5 volt) or a dim light for use us desired. Silver Plated Reflector. All Brass Front—3½ in. diameter. Takes two No. 800 Batteries and 5.5 V. Bulb.

Gives Fifty Hours Continuous Light.

No. 29.

A well designed and thoroughly efficient lamp. The complete front opens to give easy access to the battery, while the front bezel together with reflector can be removed for cleaning purposes. Diameter of Front—3½ in. Takes No. 800 Battery and 2.5 V. Bulb.

Ebony and Nickel Plated. Price **4/-** complete.

Ebony and Chromium Plated. Price **4/3** complete.

No 18 D. as illustrated, with 3-position Switch, Ebony and Chromium Plated Fittings. Price **4/9** complete.

No. 18 S.B. with Spring Back, screw-down Switch, Ebony and Nickel-Plated Fittings. Price **3/9** complete.

No. 18 as **No. 18 S.B.** with fixed back, Ebony and Nickel-Plated Fittings. Price **3/6** complete.

Diameter of Front—3½ in. Takes No. 800 Battery and 2.5 V. Bulb. A Popular Pattern Lamp with bold Brass Front, Silvered Reflector and detachable Handle.

No. 54. TAIL LAMP. **2/-** complete.

This lamp has been designed to meet the demands for a Battery Tail Lamp, and is proving itself to be a popular model with cyclists. Is fitted with a positive screw-in type Switch and Bayonet Front, which securely fastens same to body of lamp. SCREW IN SWITCH.

Fitted with 1½ in. Ruby Bi-Convex Lens. Has an adjustable clip which can be supplied for fitting to either D Stay or round forks. Takes U.2 Battery and 2.5 V. Bulb. Finish, All Ebony.

No. 15. A Lamp of Outstanding Value.

This is an improved Model which represents the best value obtainable.

The front opens to give easy access to battery, and the beaded detachable front bezel fitted gives the lamp a bold and handsome appearance.

A Parabolic Reflector is fitted, together with Spiral Filament Bulb, giving an excellent beam of light. A screw-down Switch cut thread gives perfect contact.

Diameter of Front 2½ in.

Takes No. 800 Battery and 2.5 V. Bulb.

Price **2/6** complete. All Ebony finish.

No. 101. An entirely new de-luxe Model.

This new torpedo shaped Headlamp being fitted with two green sidelights and mounted on a strong, specially designed spring bracket, gives the lamp a pleasing appearance.

Is designed to take two Flash Lamp Batteries and ample provision is made for ease of fitting.

A positive series parallel Switch gives a bright or dim light as desired.

Fitted with an 8 volt Bulb. THREE POSITION SWITCH.

Ebony with Chromium Plated Mounts.

Price **6/6** complete.

No. 102 as above but designed to take No. 801 long life Battery and fitted with 2.5 V. Bulb. Diameter of Front .. 4 in.

No. 30D as illustrated.

This lamp is fitted with a 3-position dimmer switch so that bright or dim light may be used as desired. Fitted with Sterling Silver-Plated Reflector, and All-Brass Front Nickel-Plated. Takes No. 800 Battery and 2.5 V. Bulb. Finish, Ebony and Nickel Plated. Price **3/6** complete.

No. 30, as No. 30D but with Screw-down switch. Finish Ebony and Nickel-Plated. Price **3/-** complete.

No. 31.

This Lamp is designed to take two P. & H. U.2 Batteries. Provision is made at the back of the Reflector for carrying a spare bulb. Fitted with screw-down Switch, Sterling Silver-Plated Reflector. Diameter of Front 2½ in. Takes two U.2 Batteries and 2.5 V. Bulb. Ebony and Nickel-Plated. Price **4/1** complete.

No. 39 Battery Box Set.

The Head Lamp is of pleasing design, having a 4 in. Front and Sterling Silver-Plated Reflector. The Battery Box is fitted with all-metal clips and spring grip terminals. A screw-in type Switch is used, ensuring perfect contact under all conditions. Takes P. & H. No. 800 Battery and 2.5 V. Bulb. Finish, All Ebony. Price **5/6** complete.

No. 27 Set.

As above, but with torpedo-shaped Head Lamp. Price **7/3** complete.

No. 33.

No. 33 Set, as illustrated, is an excellent Lighting Set. The Head Lamp is fitted with a change-over switch, and provision is made inside the lamp for a flash lamp battery which can be switched on when the cycle is stationary. Dynamo supplies ample current for both lamps, even when the cycle is propelled at only walking pace.

No. 33 Set complete .. **21/-**
No. 33 Set less tail lamp **19/6**
No. 28 Set, similar to above, but supplied with Headlamp as in No. 39 Set.
No. 28 Set complete .. **19/-**
No. 28 Set less tail lamp **17/6**

YOUR LIFE AND LIMB DEPEND UPON GOOD LIGHTS!

Powell & Hanmer 1932

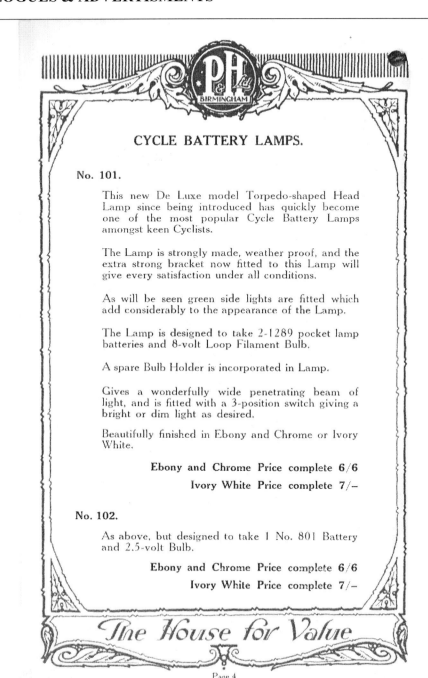

CYCLE BATTERY LAMPS.

No. 101.

This new De Luxe model Torpedo-shaped Head Lamp since being introduced has quickly become one of the most popular Cycle Battery Lamps amongst keen Cyclists.

The Lamp is strongly made, weather proof, and the extra strong bracket now fitted to this Lamp will give every satisfaction under all conditions.

As will be seen green side lights are fitted which add considerably to the appearance of the Lamp.

The Lamp is designed to take 2-1289 pocket lamp batteries and 8-volt Loop Filament Bulb.

A spare Bulb Holder is incorporated in Lamp.

Gives a wonderfully wide penetrating beam of light, and is fitted with a 3-position switch giving a bright or dim light as desired.

Beautifully finished in Ebony and Chrome or Ivory White.

Ebony and Chrome Price complete 6/6

Ivory White Price complete 7/–

No. 102.

As above, but designed to take 1 No. 801 Battery and 2.5-volt Bulb.

Ebony and Chrome Price complete 6/6

Ivory White Price complete 7/–

The House for Value

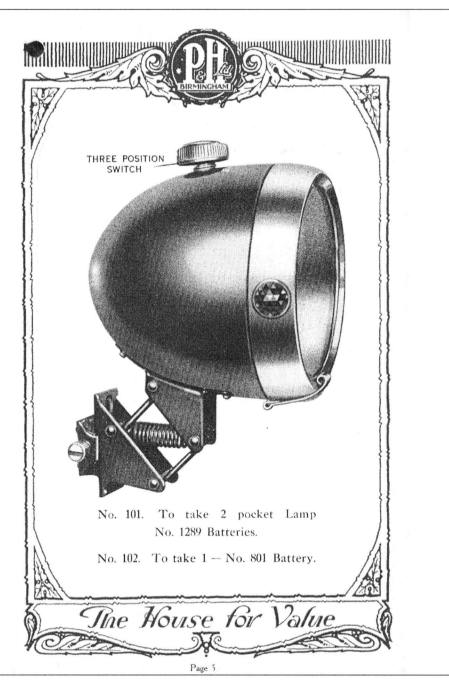

THREE POSITION SWITCH

No. 101. To take 2 pocket Lamp No. 1289 Batteries.

No. 102. To take 1 — No. 801 Battery.

The House for Value

Powell & Hanmer 1934 (1)

CYCLE BATTERY LAMPS.

This lamp has been designed to meet the needs of the rider who wishes to "Fit and Forget" for a season, and has proved itself an extremely popular model.

It is strongly made, is weatherproof, and the alternative methods of fitting lamps to cycle caters for all classes.

Provision is made for carrying two No. 800 batteries **with three-position switch to provide a full light (5.5-volt) or a dim light for use in town when desired.**

Is fitted with a Sterling Silver Plated Reflector, an all Brass Front, **and gives a wonderful wide beam of light.**

All No. 35 models are fitted with green side lights which add considerably to the appearance of same.

Diameter of front, 3¼in.; height, 4½in.

No. 35.

As will be seen from illustration, this bracket has been designed to fit on the handle bar stem, and the lamp can be fitted in this manner without fouling brake rods.

Finish: Ebony and Nickel Plated,
PRICE complete - - - 5/6

No. 35S.B.

Fitted with specially designed Spring-back, incorporating shock absorbing device.

Finish: Ebony and Nickel Plated,
PRICE complete - - - 6/-

No. 35C.P.

As No. 35S.B., but with chromium plated mounts.

Finish: Ebony and Chrome,
PRICE complete - - - 6/6

The House for Value

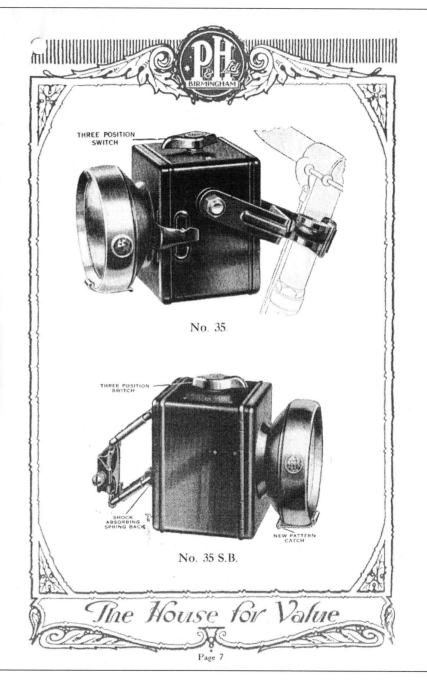

THREE POSITION SWITCH

No. 35.

THREE POSITION SWITCH

SHOCK ABSORBING SPRING BACK

NEW PATTERN CATCH

No. 35 S.B.

The House for Value

Powell & Hanmer 1934 (2)

CYCLE BATTERY LAMPS.
(Spring Back and Dimmer Models).

No. 18 range of Lamps embody many features for the discriminating cyclist. All Brass Front, complete with Bulb, Glass and Reflector fitted, is detachable as a complete unit, and is securely held by a positive fold-over catch. A Brass, Sterling Silver-Plated Reflector gives a brilliant beam of light. Positive action switch and removable handle for carrying purposes, which can be securely locked in position when used as a Cycle Lamp.

Diameter of front, 3¼in.; height, 4½in. Takes No. 800 battery and 2.5-volt bulb.

No. 18D.

Is fitted with a three-position switch, specially designed Spring Back, and the two green side lights fitted, considerably add to the appearance of the Lamp.

Finish: Ebony and Chrome.

PRICE complete - - - 4/9

No. 18S.B.

With specially designed Spring Back and positive screw-down switch.

Finish: Ebony and Nickel Plated.

PRICE complete - - - 3/9

No. 18.

With fixed bracket for fitting to cycle, and positive screw-down switch.

Finish: Ebony and Nickel Plated.

PRICE complete - - - 3/6

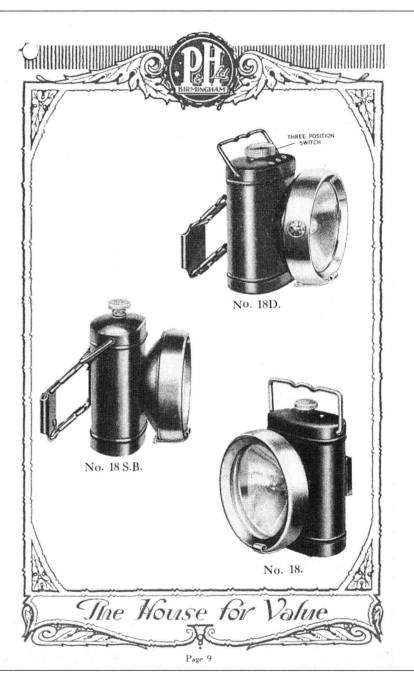

No. 18D.

No. 18 S.B.

No. 18.

Page 8

Page 9

Powell & Hanmer 1934 (3)

CYCLE BATTERY LAMPS.

No. 29.

A well-designed and thoroughly efficient lamp, the front of which opens for easy access to battery. The lamp is waterproof, and has a Sterling Silver Plated Reflector. Screw-down switch (cut thread). Can be used as Hand Inspection or Cycle Lamp. When used as a cycle lamp the handle can be locked in position at back of Lamp.

A new feature regarding this lamp is that it is fitted with a newly designed All-Brass Front, with a fold-over catch, which is simple in operation and very secure.

Diameter of Front, 3¾ in. Takes No. 800 battery and 2.5-volt bulb.

Finish: Ebony and Nickel Plated.

PRICE complete, 4/–

Finish: Ebony and Chrome.

PRICE complete, 4/3

No. 31.

This Lamp is designed to take two P. & H. U.2 Batteries.

The Lamp opens at the front to give easy access to batteries, and being fitted with a bold All-Brass Front, heavily Nickel-Plated, gives the lamp a good appearance.

Provision is made at the back of the Reflector for carrying a spare bulb. Fitted with screw-down Switch, Brass Reflector Sterling Silver-Plated.

Diameter of Front, 2¾ in. Takes two U.2 batteries and 2.5-volt bulb.

Finish: Ebony and Nickel Plated.

PRICE complete, 4/1

No. 30D. Dimmer Model.

Neatly designed Lamp, strongly made, waterproof and beautifully finished, fitted with a three-position dimmer switch so that bright or dim light may be used as desired. All Brass Front, Nickel-Plated, complete with Bulb, Glass and Reflector fitted is detachable as a complete unit, and is securely held by a positive fold-over catch. A Brass, Sterling Silver-Plated Reflector gives a brilliant beam of light. Takes No. 800 battery and 2.5-volt bulb.

Finish: Ebony and Nickel Plated.

PRICE complete, 3/6

No. 30.

As No. 30D but with a screw-down switch, takes No. 800 battery and 2.5-volt bulb.
No. 800 battery and 2.5-volt bulb.

Finish: Ebony and Nickel Plated.

PRICE complete, 3/–

No 29.

THREE POSITION SWITCH

No. 31.

No 30D.

Powell & Hanmer 1934 (4)

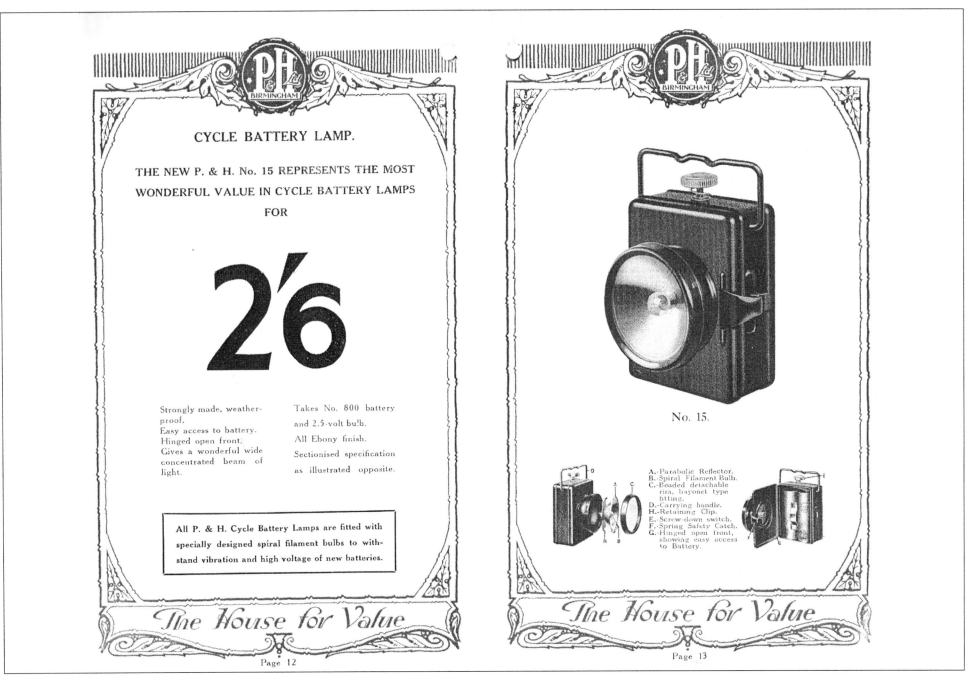

CYCLE BATTERY LAMP.

THE NEW P. & H. No. 15 REPRESENTS THE MOST WONDERFUL VALUE IN CYCLE BATTERY LAMPS

FOR

2′6

Strongly made, weatherproof.
Easy access to battery.
Hinged open front.
Gives a wonderful wide concentrated beam of light.

Takes No. 800 battery and 2.5-volt bulb.
All Ebony finish.
Sectionised specification as illustrated opposite.

All P. & H. Cycle Battery Lamps are fitted with specially designed spiral filament bulbs to withstand vibration and high voltage of new batteries.

Page 12

The House for Value

No. 15.

A.-Parabolic Reflector.
B.-Spiral Filament Bulb.
C.-Beaded detachable rim, bayonet type fitting.
D.-Carrying handle.
H.-Retaining Clip.
E.-Screw-down switch.
F.-Spring Safety Catch.
G.-Hinged open front, showing easy access to Battery.

Page 13

The House for Value

Powell & Hanmer 1934 (5)

No. 39. BATTERY BOX SET.

Those who desire this form of lighting will find this an exceptionally reliable set. The Head Lamp is of pleasing design, having a 4in. Front and Sterling Silver Plated Reflector. The Battery Box is fitted with all metal clips and spring grip terminals. A screw-in type switch is used, ensuring perfect contact under all conditions.

Takes P. & H. No. 800 Battery and 2.5-volt Bulb.

Finish: Ebony.   PRICE complete 5/6

No. 54. BATTERY TAIL LAMP.

Has been designed to meet the demands for a Battery Tail Lamp, and is proving itself to be a popular model with cyclists.

Is fitted with a positive screw-in type switch, detachable front, which is locked to body of lamp by means of a set-screw.

Fitted with 1¾in. Ruby Bi-Convex Lens.

Has an adjustable clip which can be supplied for fitting to either D Stay or Round forks.

Takes U.2 Battery and 2.5-volt Bulb.

Finish: Ebony.	**PRICE complete**	**2/-**
Finish: Ivory White.	,, ,,	**2/3**

BATTERIES.

		each.
No. 800 "Long Life" Two-cell Battery ...		8d.
No. 801 Two-cell Battery		7d.
No. U.2 Torch Unit Cell Battery ...		3½d.
No. 1289 Three-cell Pocket Lamp Battery		5d.

BULBS FOR CYCLE BATTERY LAMPS.

	each.
2.5-volt 15m/m. Spiral Filament Bulb ...	3½d.
5.5-volt Spiral Filament Bulb	6d.
8-volt .1-amp. Loop Filament Bulb ...	6d.

BULBS FOR CYCLE DYNAMO SETS.

	each.
Head Lamp 3.5-volt, .3-amp.	3½d.
Tail Lamp 4-volt .04-amp.	8d.
Head Lamp when used without Tail Lamp 4-volt .3-amp.	6d.

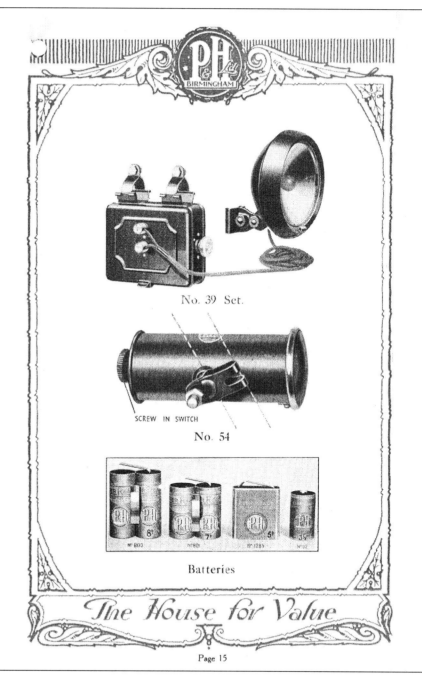

No. 39 Set.

SCREW IN SWITCH

No. 54

Batteries

Powell & Hanmer 1934 (6)

CYCLE DYNAMO LIGHTING SETS.

These sets have been designed for the rider who requires trouble-free lighting without the use of oil or carbide.

A very neat and compact dynamo supplies the current to a large front lamp and small rear lamp.

The dynamo supplies ample current for both head and rear lamps even when the cycle is propelled at only walking pace.

The Head Lamp No. 340 is fitted with a change-over switch and provision is made inside the lamp for carrying a No. 1289 flash lamp battery.

By means of the switch the rider can change over from the dynamo to the battery when the machine is stationary, i.e., in traffic hold-ups, etc.

No. 360 is similar to No. 340 except that no provision is made for carrying flash lamp battery.

No. 380 is designed for those who require a light, yet strongly-made lamp.

All Cycle Dynamo Head Lamps are fitted with green side lights and finished in Ebony with Chromium Plated front rims.

			Per Set.
No. 33 or 37 Sets, complete	...	...	20/–
No. 33 or 37 Sets, less Tail Lamp	...	18/6
No. 28 Set, complete	...	...	18/–
No. 28 Set, less Tail Lamp	...	...	16/6
			each.
No. 340 Head Lamp only	...	...	5/6
No. 360 Head Lamp only	...	...	5/6
No. 380 Head Lamp only	...	...	3/6
No. 32 Tail Lamp	...	...	1/6
Dynamo complete with Bracket	...	...	13/–

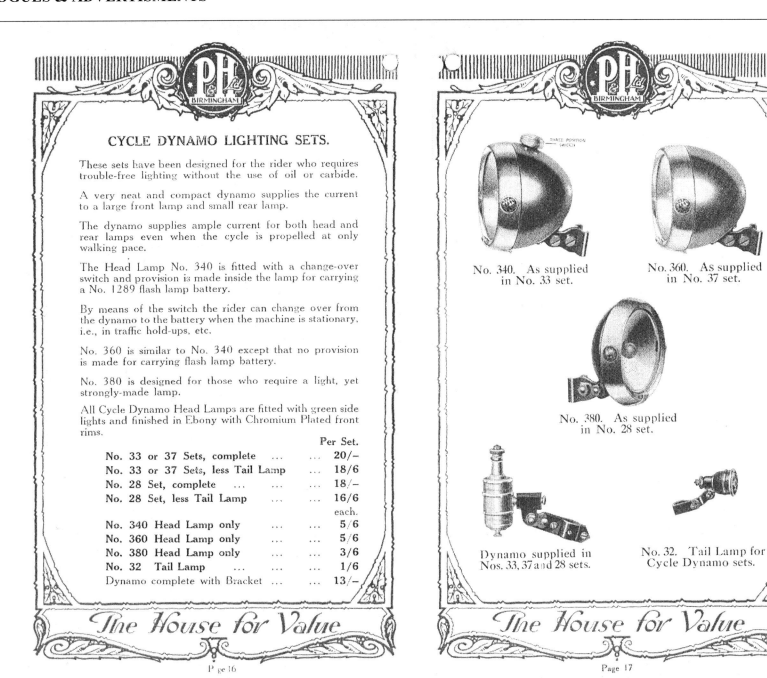

No. 340. As supplied in No. 33 set.

No. 360. As supplied in No. 37 set.

No. 380. As supplied in No. 28 set.

Dynamo supplied in Nos. 33, 37 and 28 sets.

No. 32. Tail Lamp for Cycle Dynamo sets.

The House for Value

The House for Value

Powell & Hanmer 1934 (7)

Powell & Hanmer 1936 front cover

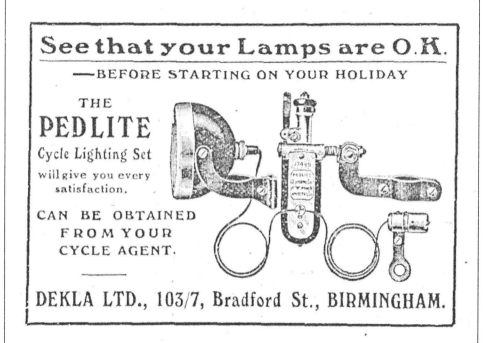

Pedlight advertisment 1922

Pifco 1939 (1)

PIFCO 6 volt SERVICE ELECTRIC LIGHTING SET

Foreign

The cheapest method of cycle lighting. Dependable though low priced. Service Lighting Set is incomparable for merit and value. Scientifically designed for the light roadster, the whole unit only weighs 30 oz. 6 volt regular light, 3 watt output. Safety in riding, regardless of speed or condition of bicycle. Instantly fitted, weatherproof and silent.

Comprises: Light-weight dynamo new style.

Head lamp with deep reflector ensuring 100 ft. beam light, with spare bulb fitted.

Tail lamp with red lens and ready fitting bracket, all 3 chromium-plated finish.

PIFCO SERVICE ELECTRIC LIGHTING SET guaranteed 12 months if branded PIFCO and packed in PIFCO printed carton, and used in accordance with printed instructions.

Avoid and refuse all substitutes.

No. 833/704]
PRICE COMPLETE

10/6

Pifco 1939 (2)

PIFCO 8 volt DE-LUXE ELECTRIC LIGHTING SET

Foreign

The most powerful 8 volt lighting unit made. Designed for the discriminating rider, weighs complete 36 oz. Instant light, front and rear, regardless of speed or condition of bicycle. Easily fitted, weatherproof, silent. Assured safety when riding.

Comprises: 8 volt dynamo, 4 watt output. Pat. Head Lamp with extra deep reflector ensuring powerful beam; spare bulb fitted. Tail Lamp, with regulation red lens and ready fitting bracket; all 3 finished chromium plated.

PIFCO DE LUXE ELECTRIC LIGHTING SET guaranteed 12 months if branded PIFCO, packed in PIFCO printed carton and used in accordance with printed instructions.

Avoid and refuse all substitutes.

No. 933/1002
PRICE COMPLETE
14/6

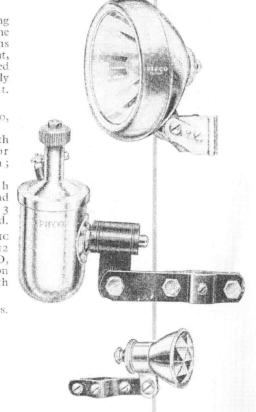

PIFCO 6 volt PREMIER ELECTRIC LIGHTING SET

Foreign

Includes features unknown in other sets of similar price. Recognised as splendid value. Comprises: Dipping head lamp, which takes pocket battery for auxiliary light when stationary. Two bulbs fitted, Dim and Bright switch to economise current. 6 volt dynamo, 3 watt output, exclusive design to PIFCO. Weatherproof. May safely be used at 20 m.p.h. speed. Tail lamp, with regulation red lens, ready fitting bracket, all 3 finished chromium plated. Complete outfit weighs 42 ozs. PIFCO PREMIER ELECTRIC LIGHTING SET. Guaranteed for 12 months, if branded PIFCO and used according to printed instructions.

Avoid and refuse all substitutes.

No. 933/1409
PRICE COMPLETE
21/-

Pifco 1939 (3)

Pifco 1939 (4)

PIFCO 6 volt NATIONAL ELECTRIC LIGHTING SET
English made

Includes more features, therefore, more attractive.
Comprises : Dipping head lamp accommodates battery for auxiliary light when not riding. Spare bulb carrier. Two bulbs and patent triangle switch for dim or bright light to economise current. Green side lights and anti-dazzle lens in chromium-plated front. 6 volt dynamo with 3 watt output. Voltage regulator. Securely enclosed, completely weatherproof. Tail lamp with regulation lens and ready fitting bracket. All finished black. Complete outfit weighs 43 ozs.

PIFCO National Electric Lighting Set is branded PIFCO and packed in PIFCO printed carton. Guaranteed for 12 months when used according to printed instructions. Avoid and refuse all substitutes.

No. 000 000
PRICE COMPLETE
23/6

Pifco 1939 (5)

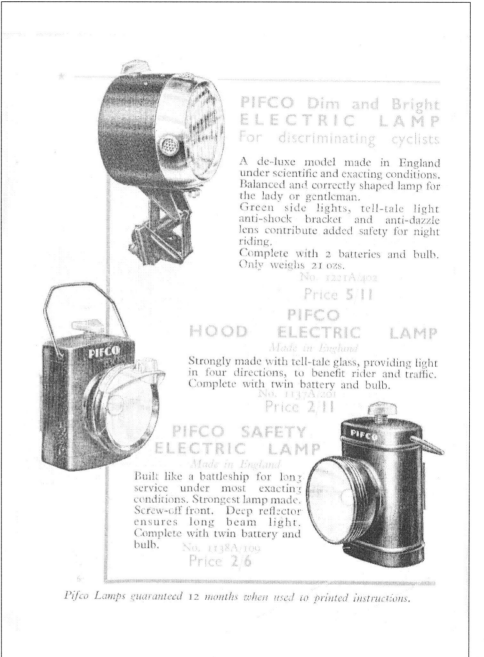

PIFCO Dim and Bright ELECTRIC LAMP
For discriminating cyclists

A de-luxe model made in England under scientific and exacting conditions. Balanced and correctly shaped lamp for the lady or gentleman. Green side lights, tell-tale light anti-shock bracket and anti-dazzle lens contribute added safety for night riding.
Complete with 2 batteries and bulb. Only weighs 21 ozs.

No. 1221A/402
Price 5/11

PIFCO HOOD ELECTRIC LAMP
Made in England

Strongly made with tell-tale glass, providing light in four directions, to benefit rider and traffic. Complete with twin battery and bulb.

No. 1137A/201
Price 2/11

PIFCO SAFETY ELECTRIC LAMP
Made in England

Built like a battleship for long service under most exacting conditions. Strongest lamp made. Screw-off front. Deep reflector ensures long beam light. Complete with twin battery and bulb.

No. 1138A/100
Price 2/6

Pifco Lamps guaranteed 12 months when used to printed instructions.

Pifco 1939 (6)

94

PIFCO ELECTRIC
RED LIGHT INDICATOR

Foreign

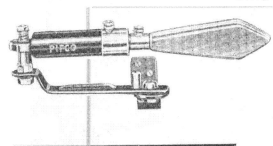

Fits all bicycles directly under expanding bolt 5-in. long left or right automatic red signal as required. A definite boon for safe riding and prevention of accidents. Complete with battery and bulb.

No. 1200A 302

Price 4 6

PIFCO MINOR
ELECTRIC TAIL LAMP, All Brass

Foreign

Nickel polished with regulation red lens. Screw-off front, deep reflector ensures long beam signal. Complete with large capacity battery and bulb.

No. 1222A 69

Price 1/–

PIFCO ELECTRIC
REAR-GUARD LAMP

English make

With Government regulation white flap. Smart safety device, clips on all rear-guards. Throws off red light signal and also illuminates the flap with bright light. Current secured from front head lamp. Supplied complete with bulb and 5-ft. flex.

No. 1223/105

Price 1/11

PIFCO CONSUL
ELECTRIC
TAIL LAMP
All Brass

Foreign

Nickel polished with regulation red lens. Screw-off front, deep reflector ensures long beam signal. Complete with large capacity battery and bulb.

No. 1224A 101½

Price 1 6

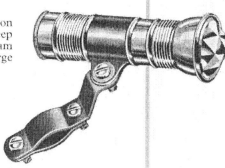

Genuine Pifco Lamps are branded and packed in Pifco Cartons.

Pifco 1939(7)

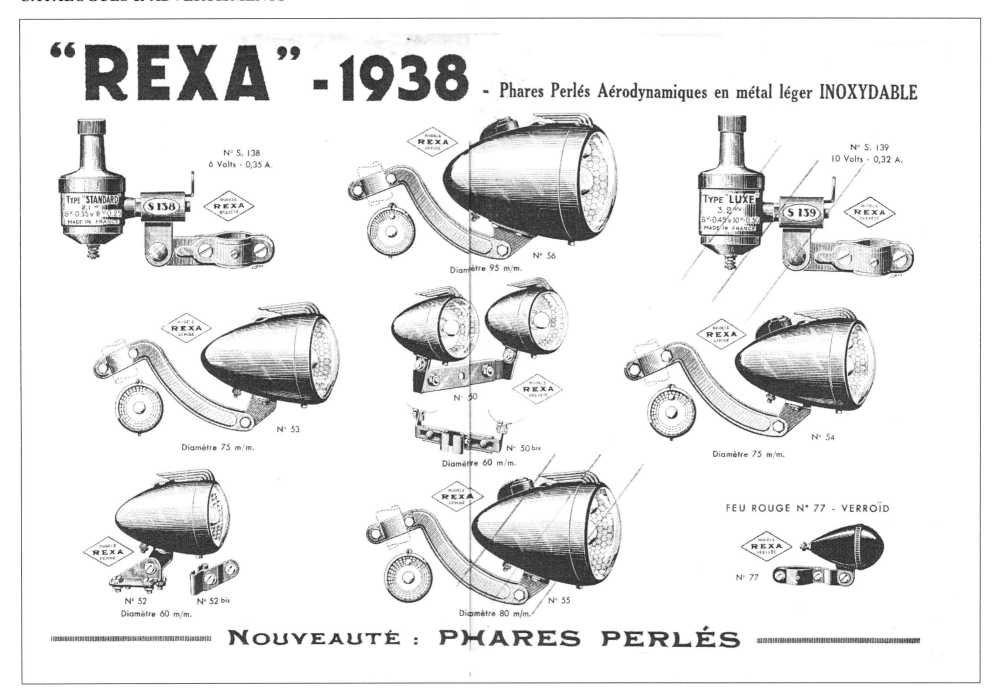

Rexa catalogue 1938

Lanterne magnéto « Riemann »

De fabrication supérieure, roulement à billes, fonctionnement garanti irréprochable, projecteur cuivre émaillé noir, verre 81 m/m, réflecteur argenté, lampe Osram 3.5 v. Poids complet., environ 715 gr.

Nº 114. — (D. 6.60, d. 31.50) .. 170 »

U. S. B. ELECTRIC BICYCLE LAMP

Follow our simple instructions and we guarantee that

"IT WORKS"

Doesn't Blow Out
or Jar Out.

Simple
Compact
Reliable

Tell your dealer to get it, or send us and we will promptly forward one of our lamps, complete.

$3.75

We also Manufacture . . . .
Electric House and Carriage Lamps.

UNITED STATES BATTERY COMPANY

253 Broadway, New York.

Recharging Device...

Recharge batteries from any direct electric current by using our simple **recharging device**, thereby furnishing light from eight to twelve hours for less than two cents.

Cheap, isn't it?

On receipt of **$1.25** additional we will forward this with the lamp, **all express charges prepaid.**

We Mail Catalogues.

H. Reimann advertisment 1925

USB Electric Lamp advertisment 1899

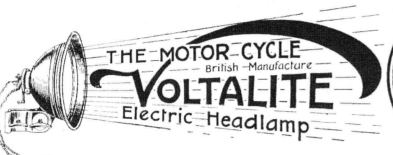

Volatlite advertisment 1912

50 Years Experience in Lamp Manufacture

H. MILLER & CO. LTD.

HEAD OFFICE & FACTORY
ASTON BROOK STREET
BIRMINGHAM 6
TELEPHONE:
ASTON CROSS 1575-6-7-8
TELEGRAMS:
MONARCH, BIRMINGHAM

LONDON OFFICE
TELEPHONE·MUSEUM 0462
CODES· BENTLEYS A.B.C (5TH EDN)
AND MOTOR TRADE CODE
CONTRACTORS TO G.P.O
H. M. WAR OFFICE R.A.F.
ADMIRALTY, MUNICIPALITIES.

GLM/LS. C.407.   30th June, 1939.

CYCLE LAMP SEASON, 1939/40.

Dear Sir(s),

Further to our letter of May 31st, when we advised you of the advance in price of several battery lamps for the coming season, we now have pleasure in enclosing our complete catalogue for 1939/40 and would particularly like to draw your attention to the following:—

DYNAMO SETS. 324RC, 324R and 325 (pages 3-5). Output increased from 6 volt 3.25 watt to 6 volt 6 watt.
537T and 526T (pages 6-7). New Ultra-lightweight sets with battery-carrying head lamps.

BATTERY LAMPS. 252 (page 14). New pattern with two bulbs and moulded lens.
242 (page 15). New pattern at popular price with 3" front.

Discount. 33⅓% off all lines excepting the Bulb Display and Spares Kit mentioned below.

To enable agents to supply promptly the correct spare bulbs and parts for MILLER products we have introduced an attractive bulb display, illustrated on page 18, and a complete cycle dynamo spares kit stocked with parts in ready demand. Net prices as follows:—

	Trade Price.	Retail Value.
Bulb Display.	£1 . 10 . 0	£2 . 6 . 0
Spares Kit.	£1 . 11 . 6	£2 . 12 . 6

Settlement Terms. 3½% for cash in seven days or 2½% monthly account.

Attractive display cards can be had on application to your factor or direct to us.

Our products are second to none in quality and design and it is our policy to continue improving them in the same way we have done in the past few years, so make MILLER products your leading lines and you will give the cyclist the best value for his money.

MILLER

MILLER CYCLE DYNAMO SET.

The Miller Cycle Dynamo Set embodies all the good points necessary to make this form of lighting successful. Full details are given in separate booklet, which will be forwarded on application.

The lamp is supplied complete with bracket, clips, and wiring, ready for attachment to bicycle, an operation which can be performed in a very few minutes.

Price :

No. 56C.D.,
Head Lamp and Dynamo,
£1 2 0

Cable Code : 00110.

No. 57C.D.,
complete with Tail Lamp,
£1 5 0

Cable Code : 00111.

The Lamps that won't go out

23

H. Miller & Co Ltd 1923

Joseph Lucas & Co Ltd Trade stand 1934